39. Colloquium der Gesellschaft für Biologische Chemie
14.–16. April 1988 in Mosbach/Baden

Protein Structure and Protein Engineering

Edited by
E.-L. Winnacker and R. Huber

With 60 Figures

Springer-Verlag Berlin Heidelberg GmbH

Prof. Dr. ERNST-LUDWIG WINNACKER
Labor für Molekularbiologie, Genzentrum, Am Klopferspitz,
8033 Martinsried, FRG

Prof. Dr. ROBERT HUBER
Max-Planck-Institut für Biochemie, Am Klopferspitz,
8033 Martinsried, FRG

ISBN 978-3-642-74175-3 ISBN 978-3-642-74173-9 (eBook)
DOI 10.1007/978-3-642-74173-9

Library of Congress Cataloging-in-Publication Data. Gesellschaft für Biologische Chemie. Colloquium (39th: 1988: Mosbach, Baden-Württemberg, Germany) Protein structure and protein engineering/edited by E.-L.Winnacker and R. Huber. p. cm. 1. Proteins-Structure-Congresses. 2. Protein engineering-Congresses. I. Winnacker, Ernst L. II. Huber, R. (Robert), 1937- . III. Title. QP551.G35 1988 574.19'245-dc 19 88-38199

Originally published by Springer-Verlag Berlin Heidelberg New York in 1988
Softcover reprint of the hardcover 1st edition 1988

2131/3130-543210

Preface

Protein engineering has had considerable impact on basic and applied research in biochemistry and molecular biology. It is already in use as a tool in molecular biology, but it is beginning to strongly influence the planning of experiments in biology everywhere, and, with even further-reaching consequences, the appointment politics in research institutions and industries.

Protein engineering, perhaps more than any other methods of protein analysis and peptide synthesis, has shown that proteins are organic molecules governed by the universal laws of chemistry and physics. However, as was the case with other new powerful methods and techniques, protein engineering tempts to an exploration of its limitations and thus generates more questions than it answers.

The 39th Mosbacher Colloquium on Protein Structure and Protein Engineering is not the first conference on this topic and it will not be the last. The important issues are obviously techniques of protein engineering, examples of application, and the basic framework of protein structure and stability within which reasonable experiments can be designed; conversely also, what we can learn about protein structure, dynamics, and folding from such experiments. Experiments in this direction aim at elucidating the folding code in the long run, but help to exploit the role of individual amino acid residues in catalysis, protein stability, and binding specificity in selected proteins now. Consequently, there are already practical applications of enzymes engineered for increased thermal stability (Alber), or of protein protease inhibitors with deliberately modified specificities (Bode). These experiments have to adhere closely to the parent molecules in structure and function to allow a rational approach on the very limited basis of our knowledge of protein structure and function, but the goal, or rather dream, is to design novel proteins with given specified functions.

This goal reaches even further than the solution of the folding code and there are doubts whether it is realistic. It may be realistic for a particular class of proteins constructed from a framework and hypervariable segments which can adopt virtually any structure, the immunoglobulins. Protein engineering with immunoglobulins follows nature, which makes use of gene rearrangement to generate virtually unlimited binding specificities. The protein engineer's approach is either to use the natural genetic repertoire and the methodology of monoclonal antibodies (Lerner) or to graft artificial amino acid sequences onto immunoglobulin frameworks (Plückthun).

In view of the many facets of protein engineering and within the given narrow time frame, the scientific program had to be quite selective and different aspects of protein structure, stability, and dynamics and different exemplary protein systems could have been chosen. The organizers nevertheless wanted to document the concepts on which experiments of protein engineering must be based and to demonstrate the kind of results expected or obtained. The experiments are rationalized in terms of protein structure so that this aspect played a central role. The forces that hold proteins in a defined three-dimensional structure result from intramolecular interactions in the polypeptide chain and from intermolecular interactions with the solvent. The study of the interplay of solute and solvent and the underlying thermodynamic parameters is a classical field, but provides new insights when modern techniques and theoretical analyses are applied (Privalov, Dill).

The protein solvent system is by far too large for a quantitative evaluation of the time-independent Schrödinger equation involving electrons and nuclei, or, derived from it, for semi-empirical and ab-initio methods of quantum chemistry. Molecular dynamics, however, which treats molecular systems as ensembles of atoms with pairwise interactions according to empirical potentials, is feasible even with present-day computing facilities and holds some promise of reproducing experimental protein structures and thermodynamic properties (Levitt). A fully empirical approach to protein structures has gained considerable importance (Jones). It uses structural data from known proteins to suggest folds of segments in unknown proteins. It is based on the observation that despite the versatility of protein structures, local geometries are conserved (canonical α-helices, β-structures, turns, etc.).

Proteins have stable structures essential for their functions but their dynamic properties seem to be equally important. Molecular motions are involved in molecular interactions at all structural levels of residues, segments, and domains (Wüthrich, Clore, Huber), not only in proteins but surprisingly also in DNA (Sigler). The model of the DNA double helix gives an impression of stiffness, which the real molecule does not have.

The problem of the pathway from the unfolded to the folded state is experimentally extremely elusive, as the intermediates are unstable and short-lived. It is even unclear whether there exists in general a unique pathway or whether renaturation proceeds by many routes (Jaenicke). Renaturation experiments of engineered protein variants add a new dimension to these studies which may finally provide an answer.

There exists an alternative to protein engineering by recombinant DNA technology: chemical synthesis. With the availability of automats and high resolution separation methods chemical synthesis has considerable advantages in the synthesis of variants by fragment condensation (Kaiser, Bayer). It has the unique possibility of introducing unnatural components and exploiting the features of such groupings (Bartlett). Automatization has also advanced DNA sequencing, the prerequisite of protein engineering. The analysis of the genomic structure of the human being is thus becoming a realistic goal (Wada).

Ultimately we wish to understand protein structure *and* function. The latter concerns the thermodynamics and kinetics of its interaction with substrate. These reactions may be very complex and involve conformational transitions, but can be analyzed in quantitative detail. The enzyme systems described by Jencks are a challenge for structural biology.

The power of protein engineering in elucidating enzyme structure and mechanism becomes obvious in a number of contributions describing results on various systems. Knowles showed that an enzyme may have evolved to a perfect catalyst where a change either has no effect or lessens its performance. He and Ferscht demonstrated that the effects of mutation on structure and function are often predictable, and thus make it possible to derive this residue's contribution quantitatively. But there are also unexpected effects, probably due to induced structural changes, which highlight our lack of understanding of protein structure. Similar results were obtained with the other systems discussed: proteases and their inhibitors (Christiansen, Bode) hemoglobin (Nagai), dihydrofolate reductase (Kraut), and phage lysozyme (Alber). It is obvious that the "unexplained" results are particularly challenging and their analysis may lead to new concepts.

Protein engineering is a further dimension in the study of protein structure and function, and provides insights which cannot be obtained by other methods. It is clear that the generation and analysis of mutants will play a decisive role in the goal of understanding structure and function of proteins in quantitative terms, should this ever be reached.

The scientific organizers had the privilege of assembling the program for an interesting theme, attractive to the speakers. The result obviously appeared interesting to many, as it attracted the largest audience that had ever attended a Mosbach Colloquium. We hope it did more than satisfy the curiosity on this rapidly developing field of protein engineering, and helped to document its potential in a realistic way.

The attraction and success of a meeting rests with the speakers. We are grateful to them for coming, many from far away. They were very cooperative and made the scientific organization a pleasure.

We thank the Gesellschaft für Biologische Chemie and its President, Prof. Dr. Oesterhelt, for giving us the opportunity to organize this colloquium, and Dr. Truscheit and Dr. Gibian and their coworkers and the local authorities, who had the burden of the technical organization of this meeting. The financial support by the Deutsche Forschungsgemeinschaft and the generous sponsorship of a number of private companies is gratefully acknowledged.

E.-L. Winnacker
R. Huber

Contents

Contributors

You will find the addresses at the beginning of the respective contribution

How Does ATP Make Work?

W. P. Jencks [1]

The question of how movement is brought about has been one of the most interesting unsolved problems in biology. Muscle contraction, active transport, and the synthesis of ATP by proton transport are coupled vectorial processes that interconvert the chemical energy of ATP and mechanical or osmotic work, often in reversible reactions. This is described by Eq. (1). The problem is to understand the meaning of the

$$ATP \rightleftharpoons ADP + Pi + work \quad (1)$$

coupling in these systems, as described by this equation. It is not obvious how to think about a chemical equation of this kind; it is not useful to compare it with a steam engine or approach it with Carnot cycles or irreversible thermodynamics. We would like to define the questions that must be answered in order that we can say that we understand the mechanism of a coupled vectorial process.

Most of the mechanisms that have been proposed for coupling in these systems have been concerned largely with binding affinities. An ion might be bound on one side of a membrane with high affinity and discharged on the other side from a low affinity site, for example. However, such mechanisms do not in themselves explain the stoichiometric hydrolysis of ATP and transport of two calcium ions against a large concentration gradient that is catalyzed by the calcium ATPase of sarcoplasmic reticulum and brings about muscle relaxation [Eq. (2)]. Binding energy is important for these systems, but it operates to avoid

$$ATP + 2\ Ca^{2+}_{out} \rightleftharpoons ADP + Pi + 2\ Ca^{2+}_{in} \quad (2)$$

unstable (or extremely stable) intermediates in the reaction path, by the use of interaction energies; it does not bring about stoichiometric coupling of ATP hydrolysis to transport or other types of movement (Jencks 1980; Pickart and Jencks 1984).

The coupling of ATP hydrolysis and work in coupled vectorial processes can be described by a set of rules. The problem of understanding coupling is then to obtain an understanding of the nature of the coupling rules. These rules generally describe enzyme specificities that change in different states of the system. The rules for the calcium-transporting ATPase, for example, can be described as follows, according to the mechanism of Eq. (3) (Pickart and Jencks 1984).

1. The enzyme species with bound Ca^{2+} catalyzes phosphoryl transfer from ATP to the enzyme and from the phosphoenzyme to ADP.

2. The enzyme species with no bound calcium catalyzes phosphoryl transfer from the phosphoenzyme to water and from inorganic phosphate to the enzyme.

[1] Graduate Department of Biochemistry, Brandeis University, Waltham, MA 02254, USA

39. Colloquium Mosbach 1988
Protein Structure and Protein Engineering

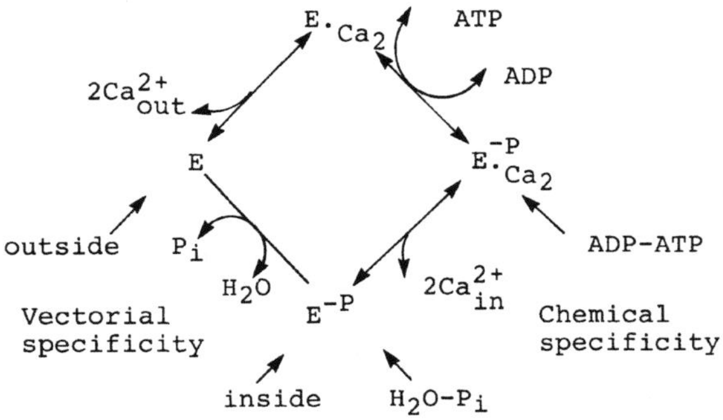

(3)

3. Calcium dissociation and binding with the free enzyme occurs only on the cytoplasmic side of the membrane.

4. Calcium binding and dissociation with the phosphoenzyme occurs only on the luminal side of the membrane.

Rules (1) and (2) are chemical specificities for catalysis of phosphoryl transfer, while rules (3) and (4) are vectorial specificities that control the movement of calcium. When these chemical and vectorial specificities are followed the system is fully coupled; to the extent that they are not followed, in any step, the reaction becomes uncoupled. There is no single step in which coupling occurs. For example, if the species $E{\sim}P \cdot Ca_2$ could react with water, there would be hydrolysis of ATP without transport and the system would be uncoupled. Similarly, if ATP could phosphorylate the free enzyme to give E-P, this would undergo hydrolysis and again uncouple the reaction. If either of the vectorial rules fail, there is leakage of calcium without ATP synthesis.

This enzyme catalyzes two reactions, the hydrolysis (or synthesis) of ATP, a chemical reaction, and the transport of two calcium ions, a vectorial reaction. The coupling rules effectively break each of these processes into two parts and insert parts of the other process in between, as shown in Eq. (4). Thus, neither the chemical nor the vectorial

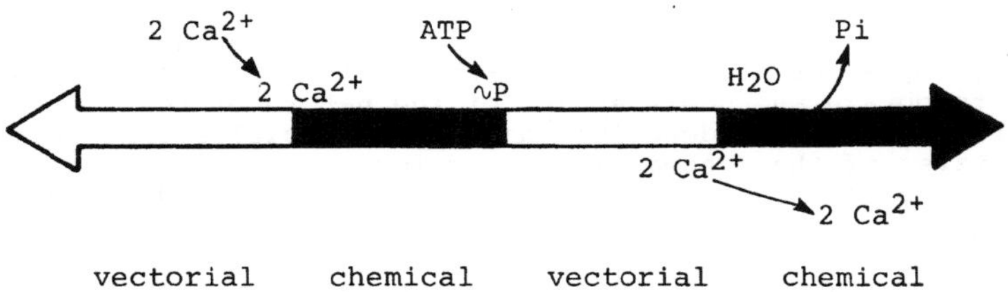

(4)

process can occur unless the other also occurs. The sodium-potassium and certain proton-transporting ATPases appear to follow the same coupling mechanism, in which Na^+ or the H^+ plays the role of Ca^{2+}.

We have been interested in how the changes in chemical specificity are related to the binding and dissociation of Ca^{2+}. In particular, it is important to know what is the specificity of enzyme species in which one Ca^{2+} ion is bound or transported.

It is well known that the free enzyme reacts readily with inorganic phosphate and the stable calcium-bound species, $^{c}E \cdot Ca_2$, reacts reversibly with ATP. A comparison of the loss of ability for phosphorylation by ATP with the rates for dissociation of the two Ca^{2+} ions from ${}^{c}E{\cdot}Ca_2$ has shown that this ability to react with ATP is lost when one Ca^{2+} dissociates; i.e., the species ${}^{c}E{\cdot}Ca_1$ is unreactive toward ATP (Petithory and Jencks 1988). This is what is desired for the function of the enzyme because the stoichiometry for transport of 2 Ca^{2+} in Eqs. (2) and (3) would not be observed if ${}^{c}E{\cdot}Ca_1$ were phosphorylated.

The second vectorial half-reaction involves the internalization of 2 Ca^{2+} from the species $E{\sim}P{\cdot}Ca_2$. This occurs after phosphorylation of the enzyme by ATP, which changes the vectorial specificity of the enzyme so that Ca^{2+} will dissociate only into the interior of the vesicle, not the cytoplasm. It is known that the two Ca^{2+} ions are also dissociated sequentially into the interior of the vesicle and that the Ca^{2+} that adds first from the cytoplasm is the ion that dissociates first from $E{\sim}P{\cdot}Ca_2$ into the lumen of the vesicle (Inesi 1987). In order that two Ca^{2+} should be internalized, the change in catalytic specificity of the enzyme from reaction with ADP-ATP to reaction with P_i-H_2O should occur only after both Ca^{2+} ions have dissociated inside; if hydrolysis were to occur after the transport of only one Ca^{2+} the stoichiometry for transport of two Ca^{2+} ions might not be achieved.

Both of these reaction sequences appear to be reasonable in order to account for the coupled transport of 2 Ca^{2+}, but they pose an apparent paradox for the species with one bound Ca^{2+} ion: the unphosphorylated enzyme changes its specificity so that it no longer reacts with ATP when a single Ca^{2+} dissociates from it, while the phosphorylated enzyme does not change its specificity until both Ca^{2+} ions are dissociated and it reacts rapidly with water. At first glance, this situation seems to require that the enzyme has different specificities for the same reaction in the two directions.

The paradox may be resolved by an earlier finding that the phosphorylation by ATP occurs in two steps. In the first step the enzyme with the two bound Ca^{2+} undergoes a conformational change, with a rate constant of 220/s, to form a catalytically active species, ${}^{a}E{\cdot}Ca_2{\cdot}ATP$. This species undergoes very rapid phosphorylation, with a rate constant of >1000/s, in the second step [Eq. (5)] (Petithory and Jencks 1986). The inability of the species with one bound calcium ion, ${}^{c}E{\cdot}Ca_1$, to become

$$
\begin{array}{ccccc}
ATP + {}^{c}E{\cdot}Ca_2 & \xrightarrow{220\ s^{-1}} & {}^{a}E{\cdot}Ca_2{\cdot}ATP & \xleftrightarrow{>1000\ s^{-1}} & E{\sim}P{\cdot}Ca_2 + ADP \\
\updownarrow & & & & \updownarrow \\
ATP + {}^{c}E{\cdot}Ca_1 & \not\longleftrightarrow & {}^{a}E{\cdot}Ca_1{\cdot}ATP & \longleftrightarrow & E{\sim}P{\cdot}Ca_1 + ADP \\
\updownarrow & & & & \updownarrow \\
P_i + E & & \longleftrightarrow & & E\text{-}P + H_2O
\end{array}
\qquad (5)
$$

phosphorylated indicates that it cannot undergo the conformational change to the catalytically active form. However, in the reverse direction, the phosphoenzyme can maintain its reactivity toward ADP, because it is in the active conformation, and changes its specificity to catalysis of phosphoenzyme hydrolysis only after both Ca^{2+} ions have dissociated to the inside of the vesicle.

We were initially puzzled as to why the enzyme followed this two-step mechanism for phosphorylation. The different requirements for specificity in the two half reactions for the transport of two Ca^{2+} ions provides a rationale for the two-step mechanism and ensures that the stoichiometry of calcium transport will be two so long as the specificities are maintained.

Other Coupled Vectorial Processes

The coupling rules are less well understood for other ATP-coupled ion pumps, but it may be useful to outline the kinds of mechanisms that may be involved. It is now well established that the proton ATPase, which is responsible for the synthesis of ATP in oxidative phosphorylation and photophosphorylation, will catalyze the reversible synthesis of ATP at the active site with no input of energy (Stroop and Boyer 1985; Penefsky 1985). The problem is to understand how the enzyme avoids the uncoupled hydrolysis of ATP and is able to couple ATP synthesis to the transport of protons.

This can occur if the chemical and vectorial processes are each broken into two parts that are separated by a step of the alternative process, as shown in Eq. (6). The chemical process is broken into (a) the reversible hydrolysis and synthesis of ATP at the active site, which is

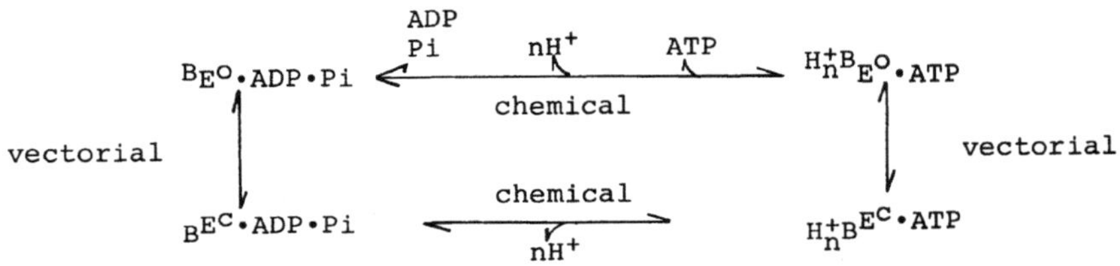

(6)

catalyzed only by the "closed" form of the enzyme, E^c, and (b) the binding and dissociation of phosphate compounds, which occurs only with the open form of the enzyme, E^o. The system will be coupled if these chemical steps are separated by two vectorial steps, such as a conformational change that exposes the empty proton-carrying site alternately to the two sides of the membrane with E•ADP•Pi, and a conformational change that exposes the occupied proton-carrying site alternately to the inside and outside with E•ATP. If these specificity rules are followed, proton transport in the two directions will be stoichiometrically coupled to ATP synthesis or hydrolysis (Jencks 1988). There is no reason to think that this particular mechanism for coupling is correct, but this or other models might stimulate the design of experiments to determine what is the correct model.

Many of the steps in the contraction of striated muscle are now understood. These steps provide an alternation of chemical and vectorial processes that are responsible for the coupling of ATP hydrolysis to the performance of mechanical work (Jencks 1986, 1988). However, the mechanism of coupling for the critical vectorial rule is not understood. This rule requires that the myosin-products complex, M•ADP•Pi, must combine with an actin monomer in the thin filament that is different from the site from which M•ADP dissociated. This is necessary in order that the two filaments should move relative to each other and carry out work. There are three possible mechanisms for this coupling rule:

1. The hydrolysis of bound ADP to give M•ADP•Pi brings about a conformational change that moves the myosin head so that it cannot recombine with the original actin monomer, but is able to combine with a different monomer on the actin filament (Lymn and Taylor 1971).

2. Binding of ATP to actomyosin brings about a conformational change that causes dissociation and also movement of the myosin head so that recombination of the myosin and bound nucleotide occurs at a different actin monomer (Lymn and Taylor 1971; Eisenberg and Hill 1985).

3. Movement of the dissociated myosin from one actin monomer to another is brought about by movement of the thick filament itself, which occurs while the actin and myosin are dissociated. The power stroke of other myosin heads moves the filament enough so that recombination of the myosin-products complex occurs with an actin monomer different from the site at which dissociation occurred (Jencks 1986).

Finally, the light-mediated proton transfer catalyzed by bacteriorhodopsin deserves mention. It is clear that the light reaction brings about an isomerization of the protonated retinal imine that changes the side of the membrane from which the proton can dissociate. What is not so clear is the nature of the coupling rule that prevents the protonated carrier from reverting back to the starting conformation in the dark reaction. If there were not such a coupling rule, net proton transfer would not occur against an unfavorable proton gradient.

Two mechanisms have recently been proposed for the coupling rule in the dark reaction. According to one mechanism, the unprotonated carrier undergoes rapid rotation about a single bond that changes the side of the membrane to which the imine nitrogen ion is exposed. The protonated carrier will undergo this rotation some 10^5 more slowly because there is less bond alternation and more double bond character in this bond when the nitrogen atom is protonated (Schulten et al. 1984). According to the other interpretation, the change of exposure is brought about simply by inversion of the imine nitrogen atom, so that its electron pair becomes exposed to the other side of the membrane. This can occur only when the nitrogen atom is unprotonated (Smith et al. 1986).

It should be evident from this summary that much remains to be investigated in order to obtain an understanding of the nature of the coupling rules and the mechanisms by which these rules are enforced in coupled vectorial processes. However, considerable progress has been made at the present in a number of different laboratories and we are now beginning to understand at least the questions that must be asked, and answered, if we are to understand the mechanism of Eq. (1).

References

Eisenberg E, Hill TL (1985) Science 277:999-1006

Inesi G (1987) J Biol Chem 262:16338-16342

Jencks WP (1980) Adv Enzymol 51:75-106

Jencks WP (1986) Methods Enzymol (in press)

Jencks WP (1988) In: Kleinkauf, von Döhren, Jaenicke (eds) The Roots of Modern Biochemistry. de Gruyter, Berlin, pp 569-580

Lymn RW, Taylor EW (1971) Biochemistry 10:4617-4624

Penefsky HS (1985) J Biol Chem 260:13735-13741

Petithory JR, Jencks WP (1986) Biochemistry 25:4493-4497

Petithory JR, Jencks WP (1988) Biochemistry 27:5553-5564

Pickart CM, Jencks WP (1984) J Biol Chem 259:1629-1643

Schulten K, Schulten Z, Tavan P (1984) In: Information and Energy Transduction in Biological Membranes, Liss, New York, pp 113-131

Smith SO, Hornung I, van der Steen R, Pardoen JA, Braiman MS, Lugtenburg J, Mathies RA (1986) Proc Natl Acad Sci USA 83:967-971

Stroop SD, Boyer PD (1985) Biochemistry 24:2304-2310

Hydrophobic Interactions in Proteins

P. L. Privalov [1]

In discussions on the mechanism of the formation and stabilization of the compact structure of native protein, the hydrophobic interactions are at present attracting the greatest attention of all the various interactions existing between the groups of these macromolecules. "Hydrophobic interaction" is usually understood as the cause of low solubility of nonpolar substances in water, of their hydrophobicity, which results in an enhanced association of these solutes in aqueous solutions. As there are many nonpolar groups in proteins and many of them are clustered together, as if avoiding contact with water, one can suppose that the hydrophobicity of these groups plays an important role in determining the compact state of globular protein just as it does in the case of an oil drop in water.

The heightened interest in hydrophobic interactions arises also from their unusual character: they appear to be driven not by the enthalpic but by the entropic features of the thermodynamic potential (see Frank and Evans 1945; Kauzmann 1959; Tanford 1980). This conclusion has been drawn from studies of dissolution of nonpolar substances in water.

Thermodynamically, the poor solubility, X, of any solute means that its transfer into the solvent requires a great expenditure, i.e., the Gibbs energy of its transfer, ΔG, is large and positive since

$$\Delta G = -RT \ln X, \quad (1)$$

and X, which is usually expressed in molar fractions, is small.

The most surprising finding was that the poor solubility of nonpolar substances in water is minimal at room temperature, i.e., at this temperature these substances are most hydrophobic (Fig. 1). Mathematically, this means that at this temperature T_H:

$$[(\partial \ln X)/(\partial T)]_{T_H} = -\{[\partial(\Delta G/RT)]/\partial T\}_{T_H} = (\Delta H)/RT^2 = 0, \quad (2)$$

i.e., the enthalpy of transfer of a nonpolar substance from the pure liquid phase to water, ΔH, is zero. And since

$$\Delta G = \Delta H - T\Delta S, \quad (3)$$

one comes to the conclusion that the large Gibbs energy of transfer of a nonpolar solute to water, the hydrophobicity of the solute, is caused by the entropy of transfer, which is large and negative (Table 1).

This thermodynamically unfavorable entropy decrease upon transfer of a nonpolar substance to water was explained by the ordering of water in the presence of this substance. One of the arguments for such an

[1] Institute of Protein Research, Academy of Sciences of the USSR, Pushchino, Moscow Region, 142292, USSR

39. Colloquium Mosbach 1988
Protein Structure and Protein Engineering

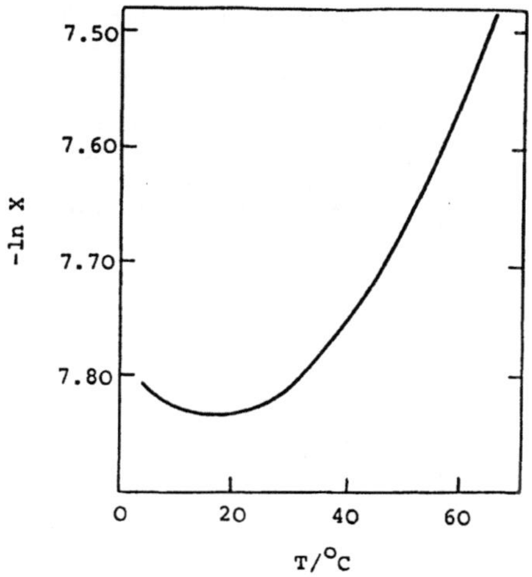

Fig. 1. Temperature dependence of solubility of benzene in water (Franks et al. 1959)

Table 1. Solubility, Gibbs energy, enthalpy, entropy and heat capacity increment of transfer of some typical nonpolar substances from the pure liquid phase to water at 25°C[a]

Substance and its surface area in Å^2	Solubility in mole fractions	ΔG in kJ/mol	ΔH in kJ/mol	ΔS in J/K•mol	ΔC_p in J/K•mol
Benzene 240	4.01×10^{-4}	19.4	2.08	-58	225
Toluene 275	1.01×10^{-4}	22.8	1.73	-71	263
Ethylbenzene 302	0.258×10^{-4}	26.2	2.02	-81	318

[a]For references, see Privalov and Gill (1988)

assumption was the observation that the entropy decrease upon the transfer of a nonpolar molecule to water is proportional to the surface area of this molecule (Table 1). This agreed with the finding that dissolution of nonpolar substances in water results in a considerable heat capacity increase, proportional to the surface area of the solute molecules (Gill and Wadso 1976). The heat capacity increment is what can be expected if a temperature increase leads to a gradual melting of water, ordered by the presence of nonpolar molecules. If so, the heat capacity increment should decrease at increasing temperature, as the influence of a nonpolar substance on the state of water should vanish at elevated temperature (Gill et al. 1985). This was found to be true in calorimetric studies of the heat capacity of solutes in an aqueous solution of nonpolar substances (Fig. 2).

A similar effect of heat capacity increase was found at the unfolding of a compact protein structure, the denaturation of protein.

The denaturation of globular protein caused by any means (denaturants, acids, or temperature) results in a significant heat capacity increment

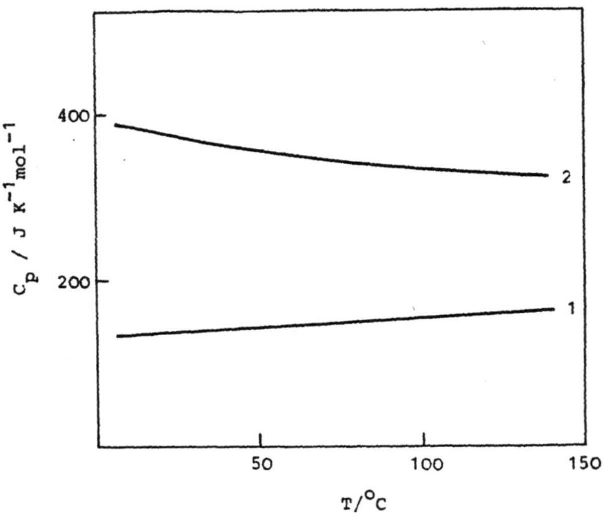

Fig. 2. Temperature dependence of the heat capacity of liquid benzene *(1)* and of its partial heat capacity *(2)* in an aqueous solution. (For details see Makhatadze and Privalov 1988)

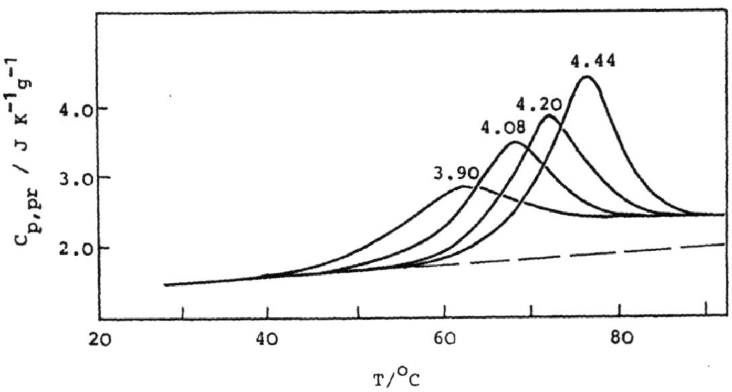

Fig. 3. Temperature dependence of the partial heat capacity of myoglobin in aqueous solutions with different pH's. The heat capacity peak corresponds to denaturational heat absorption

(Fig. 3) which does not depend on the environmental condition, but is very specific for a given protein (Privalov 1979). Its value largely exceeds that expected for the increase in vibrational freedom of a polypeptide chain at the breakdown of the rigid native structure (Sturtevant 1977) and shows a clear proportionality with the number of contacts between the nonpolar groups in native protein. Earlier, the denaturational heat capacity increment was supposed to be independent of temperature, but our recent calorimetric studies, carried out in a broader temperature range, showed that it decreases with a temperature increase in almost the same way as the heat capacity increment of transfer of a nonpolar solute to water (Fig. 4). Therefore, there is no doubt that the observed denaturational heat capacity increment is caused mainly by the hydration of the nonpolar groups, which are exposed to water upon unfolding of a compact protein structure (Kauzmann 1959; Brandts 1964; Sturtevant 1977; Privalov 1979).

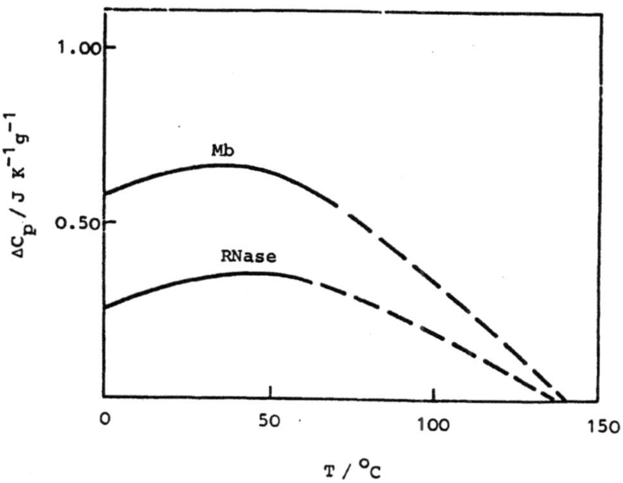

Fig. 4. Temperature dependence of the denaturational heat capacity increment of myoglobin *(Mb)* and pancreatic ribonuclease A *(RNase)* (for details see Privalov et al. 1988)

The denaturational heat capacity increment was one of the main arguments for extrapolating the entropic conception of hydrophobicity of nonpolar solutes on proteins. After Kauzmann (1959), it became conventional among protein chemists to regard the negative entropy of hydration of nonpolar groups as a cause of an independent type of molecular force, the hydrophobic interaction. It was believed that they could explain the stability of a compact state of the protein molecule, and nobody worried whether one could extrapolate the conclusions, made at a very specific temperature of maximal hydrophobicity of a nonpolar substance, to any other temperature.

At the same time, as is evident, the heat capacity increment of denaturation and transfer of a nonpolar solute to water means not only the existence of an extra order in water, but that this order decreases on temperature increase. Correspondingly, heating should lead to a decrease of the absolute value of the negative entropy of transfer of a nonpolar solute to water since

$$\Delta S(T) = \Delta S(T_o) + \int_{T_o}^{T} \Delta C_p \, d \ln T. \quad (4)$$

It is interesting that the temperature T_s, at which the entropy of transfer of a nonpolar solute to water, ΔS, becomes zero, was found to be universal for all the studied substances being about 140°C (Fig. 5) (Baldwin 1987; Privalov and Gill 1988).

The temperature increase should lead also to an enthalpy increase since

$$\Delta H(T) = \Delta H(T_o) + \int_{T_o}^{T} \Delta C_p \, dT. \quad (5)$$

If the enthalpy of transfer of a nonpolar substance from the pure liquid phase to water is zero at room temperature, at 140°C it becomes large and positive, almost of the same order of magnitude as the enthalpy of vaporization of the liquid phase corrected for the volume

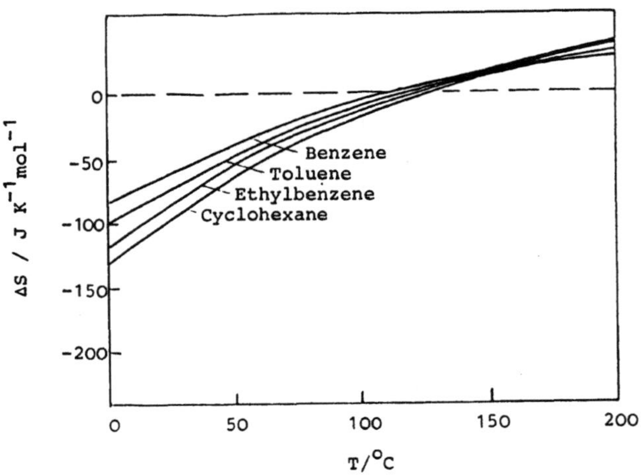

Fig. 5. Temperature dependence of the entropy of transfer of various nonpolar substances from the pure liquid phase to water

expansion, i.e., close to the energy of van der Waals interaction between nonpolar molecules.

It follows then that at this temperature T_S, 140°C, water does not solvate the nonpolar solute.

With proteins the situation is similar. Inasmuch as the denaturation heat capacity increment is positive, the difference enthalpy and difference entropy of the native and denatured states of protein are increasing functions of temperature. As shown in Figs. 6 and 7, at an increasing temperature they both approach asymptotically some constant values at 140°C (Privalov 1979; Privalov and Gill 1988). The meaning of these values is clear, as 140°C is the temperature at which water does not solvate nonpolar groups. Therefore, the enthalpy and entropy of protein denaturation at 140°C do not include the hydration effects but correspond to the conformational transition of the polypeptide chain with a disruption of intramolecular bonds (primarily hydrogen and van der Waals) maintaining the compact protein structure.

The principal difference between the two considered processes, the transfer of a nonpolar solute from the nonpolar environment to water and the denaturation of protein, is in the entropy values: while the entropy of transfer of a nonpolar solute to water is zero at 140°C, the entropy of the protein denaturation at this temperature is large and positive.

The difference in the entropy value at a temperature where hydration effects are absent might only mean that the protein interior is not a liquid-like nonpolar phase but is a crystal-like phase, which is specified by a definite positive melting entropy. The main thermodynamic consequence of this specificity of a protein molecule is that the entropy of protein transition from the native to a denatured state becomes zero at a much lower temperature than the entropy of transfer of nonpolar substances from the liquid phase to water. Usually the temperature $T_S{}^d$ for the protein denaturation is only few degrees higher than the temperature $T_H{}^d$, at which the enthalpy of protein denaturation becomes zero. For most of the known proteins, this temperature is between 0 and 30°C (Privalov 1979).

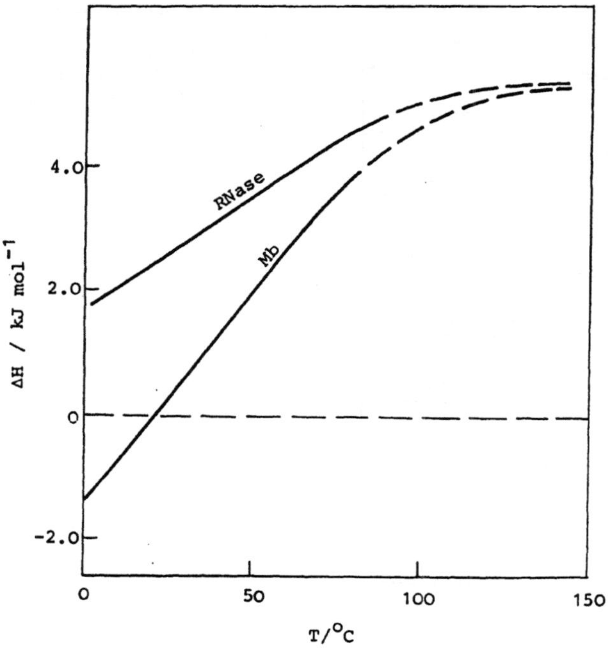

Fig. 6. Temperature dependence of the difference enthalpy of the native and denatured states for myoglobin *(Mb)* and ribonuclease *(RNase)* (Privalov et al. 1988)

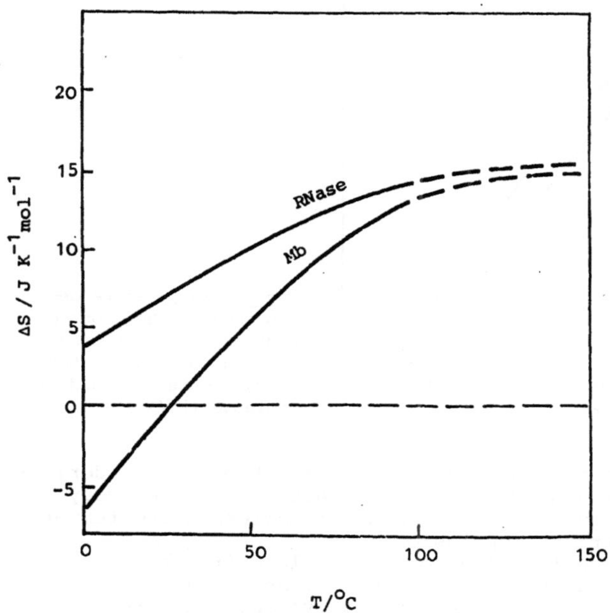

Fig. 7. Temperature dependence of the difference entropy of the native and denatured states for myoglobin *(Mb)* and ribonuclease *(RNase)*

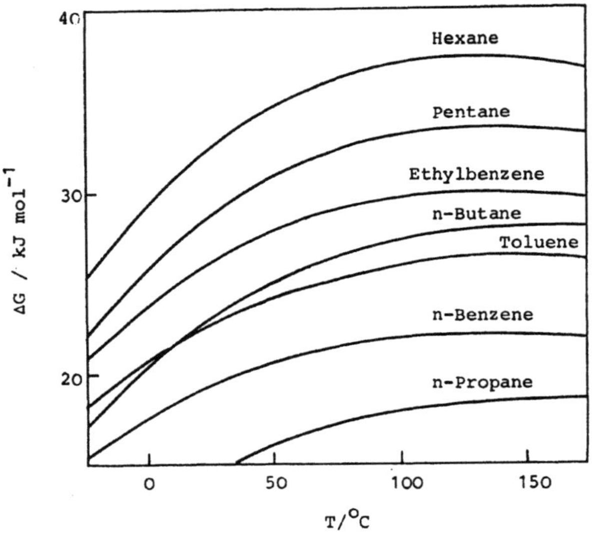

Fig. 8. Temperature dependence of the Gibbs energy of transfer for various liquid nonpolar substances to water

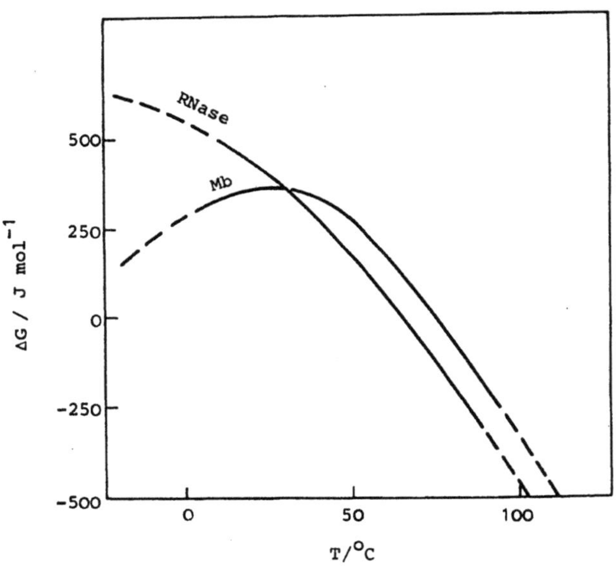

Fig. 9. Temperature dependence of the Gibbs energy difference of the native and denatured states of myoglobin *(Mb)* and ribonuclease *(RNAse)*

Now, if we take into account that the entropy is a temperature derivative of the Gibbs energy, $\partial \Delta G/\partial T = - \Delta S$, and that the zero entropy means that the Gibbs energy has the extremum at corresponding temperature, we come to two important conclusions:

a) The Gibbs energy of transfer of a nonpolar solute from the pure liquid phase to water is maximal at T_s, i.e., at about 140°C (Fig. 8).

b) The Gibbs energy difference between the native and denatured states of protein, which determines the stability of its native structure, is maximal at $T_s{}^d$, i.e., at a temperature between 0 and 30°C for most globular proteins (Fig. 9).

The Gibbs energy of transfer of a nonpolar solute from the pure liquid phase to water represents actually a measure of the expelling action of water on this solute, i.e., a measure of hydrophobic interaction. As we see, its maximal value is reached at a temperature where water does not solvate the nonpolar solute and the whole value of ΔG is provided by the van der Waals interactions between the nonpolar molecules. The decrease of ΔG from its maximal value at decreasing temperature appears simultaneously with the appearance of signs of water solvation by a nonpolar solute: the negative entropy and lower enthalpy of transfer and the significant heat capacity increment. Without hydration, without the heat capacity increment caused by water solvation, the Gibbs energy of transfer would not change at a temperature decrease from T_s. This means that the Gibbs energy of hydration of a nonpolar solute is negative: it equals zero at T_s and increases in the absolute value at a decreasing temperature. It follows from the above that the hydrophobic interaction is provided by the van der Waals interactions between the nonpolar molecules, while the hydration of these molecules has only a negative effect and it increases the solubility of these solutes in water (Shinoda 1977; Hvidt 1983; Privalov and Gill 1988).

The principal difference between the Gibbs energy of transfer of a nonpolar solute to water and the Gibbs energy of protein denaturation is caused by the entropy of conformational transition of protein which is positive and large. If protein denaturation proceeded without hydration of nonpolar groups, i.e., without a heat capacity increase, the enthalpy and entropy of protein denaturation would not depend on temperature, being ΔH^o and ΔS^o, respectively, and the Gibbs energy difference of the native and denatured states ($\Delta G = \Delta H^o - T\Delta S^o$) would be a linearly decreasing function of temperature, because the dissipative force of the thermal motion ($T\Delta S^o$) is proportional to temperature (Fig. 10). However, when protein denaturation proceeds with a heat capacity increase, i.e., with hydration of the nonpolar groups exposed upon folding, then the Gibbs energy difference of the native and dena-

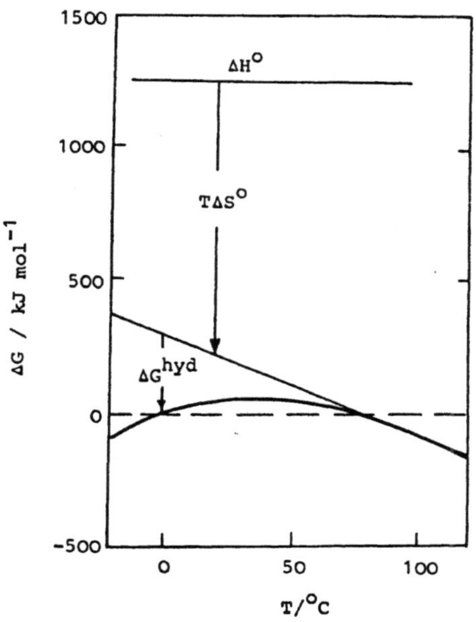

Fig. 10. Contribution of the dissipative force $T\Delta S^o$, and water solvation effect, ΔG^{hyd}, to the stabilization of an abstract globular protein consisting of about 200 amino acid residues. (For details see Privalov 1982; Privalov and Gill 1988)

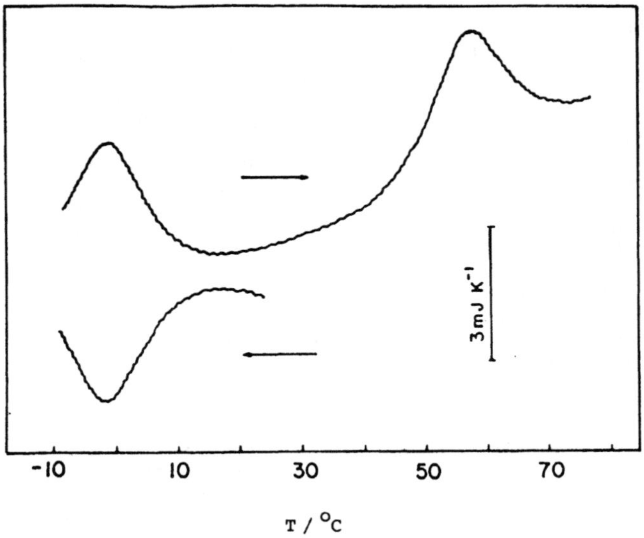

Fig. 11. Microcalorimetric recording of the heat effect upon cooling and subsequent heating of an aqueous solution of apomyoglobin (pH 4.8). The low temperature peaks correspond to the heat release upon cold denaturation and to the heat absorption upon renaturation of the protein at subsequent heating; the high temperature peak corresponds to the heat absorption upon heat denaturation. (For details see Griko et al. 1988)

tured states deviates from the linear course. The correction of the Gibbs energy function provided by hydration, ΔG^{hyd}, is always negative and increases in the absolute value at a decreasing temperature. As a result, at some sufficiently low temperature, the stability of the protein native state can decrease to zero and the protein denatures (Privalov 1982).

Cold denaturation of protein has been demonstrated on a number of globular proteins (see, e.g., Privalov et al. 1986; Griko et al. 1988). In contrast to heat denaturation, cold denaturation of proteins proceeds with a release of heat, i.e., with a decrease of the enthalpy and entropy, showing that it is accompanied by extensive hydration of the nonpolar groups exposed upon denaturation (Fig. 11).

Thus, it emerges that hydration of nonpolar groups, in spite of the widely held opinion, is a thermodynamically favorable process which can only decrease the hydrophobicity of nonpolar solutes and destabilize the compact protein state. This raises the question whether the use of the concept of "hydrophobic interaction" is justified in considering the protein stability problem. One can find a detailed consideration of this problem in the recent paper of Privalov and Gill (1988).

References

Baldwin RL (1987) Temperature dependence of the hydrophobic interaction in protein folding. Proc Natl Acad Sci USA 83:8069-8072

Brandts JF (1964) The thermodynamics of protein denaturation. I. The denaturation of chymotrypsin. J Am Chem Soc 86:4291-4301

Franks HS, Evans MW (1945) Free volume and entropy in condensed systems. J Chem Phys 11:507-532

Franks F, Gent M, Johnson HH (1963) The solubility of benzene in water. J Chem.Soc 8:2716-2723
Gill SJ, Wadso I (1976) An equation of state describing hydrophobic interactions. Proc Natl Acad Sci USA 73:2955-2958
Gill SJ, Nichols NF, Wadso I (1976) Calorimetric determination of enthalpies of solution of slightly soluble liquids. J Chem Thermodynamics 8:445-452
Gill SJ, Dec SF, Olofsson G, Wadso I (1985) Anomalous heat capacity of hydrophobic solvation. J Phys Chem 89:3758-3761
Griko YV, Privalov PL, Venyaminov SY, Kutyshenko VP (1988) Thermodynamic study of the apomyoglobin structure. J Mol Biol 202:127-138
Hvidt A (1983) The solubility of hydrocarbons in water. Acta Chem Scand 37:99-103
Kauzmann W (1959) Some factors in the interpretation of protein denaturation. Adv Protein Chem 14:1-63
Makhatadze GI, Privalov PL (1988) Partial heat capacity of benzene and toluene in aqueous solutions determined calorimetrically in a broad temperature range. J Chem Thermodynamics 20:405-412
Privalov PL (1979) Stability of proteins. Small compact globular proteins. Adv Protein Chem 33:167-241
Privalov PL (1982) Stability of proteins. Proteins which do not present a single cooperative system. Adv Protein Chem 35:1-104
Privalov PL, Griko YV, Venyaminov SY, Kutyshenko VP (1986) Cold denaturation of myoglobin. J Mol Biol 190:487-498
Privalov PL, Gill SJ (1988) Stability of protein structure and hydrophobic interactions. Adv Protein Chem (in press)
Privalov PL, Tiktopulo EI, Venyaminov SY, Griko YV, Makhatadze GI, Khechinashvili NN (1988) Heat capacity and conformation of proteins in the denatured state. J Mol Biol (in press)
Shinoda K (1977) Iceberg formation and solubility. J Phys Chem 81:1300-1302
Sturtevant JM (1977) Heat capacity and entropy changes in processes involving proteins. Proc Natl Acad Sci USA 74:2236-2240
Tanford C (1980) The hydrophobic effect: formation of micelles and biological membrane. Wiley Interscience, New York

Is There a Code for Protein Folding?*

R. Jaenicke [1]

Introduction

According to the "central dogma of molecular biology", genes encode proteins, based on the colinear relationship defined by the genetic code. Elucidating protein sequences by sequencing DNA has been proving the validity of the dogma over many years.

The detailed mechanism of protein biosynthesis, i.e., the translation of the one-dimensional information laid down in the sequence of nucleotides into the one-dimensional information at the level of the amino acid sequence, is well established. Although highly complex, each piece in the puzzle of DNA, mRNA, AA-tRNA, and the various initiation, elongation, and termination factors has been put into its proper place. However, the real importance of a gene is still obscure. It will remain so until we succeed in cracking the "second half of the genetic code" determining the translation of the one-dimensional information of the specific amino acid sequence into the corresponding unique three-dimensional arrangement of its polypeptide chain. If this "folding code" of proteins were known, it would not only aid in engineered changes in protein structure using recombinant DNA techniques, but it would also provide a tool to specify hypothetical functions for given amino-acid sequences. At present, the code of protein folding is totally unknown, and there is no hint how it might be deciphered, even for small single-chain proteins. In the case of long polypeptide chains and oligomeric proteins, domains and subunit interactions come into the play. Evidently they complicate the problem due to their specific distribution of hydrophobic residues in clusters in the interior and in patches in the intersubunit interfaces.

Fundamentals of Structure

In developing a general theory of the structure of biological systems, one has to describe how the intrinsic physical and chemical properties of the building blocks and their mutual interactions combine to determine (1) the energetics of the three-dimensional array of matter, (2) the kinetics of its formation and (3) the mechanism underlying the transition from disorder to order. In the case of biological systems, evolution has been optimizing the correlation of structure and function. Kepler (1611) was the first to point to the fact that in considering biological structure it is essential to keep in mind that

*Dedicated to Professor Theodor Wieland on occasion of his 75th birthday.

[1] Institut für Biophysik und Physikalische Biochemie, Universität Regensburg, Universitätsstraße 31, D-8400 Regensburg, FRG

39. Colloquium Mosbach 1988
Protein Structure and Protein Engineering

Table 1. Number of combinations (N) of 2n SH groups to form disulfide bonds in proteins. (Anfinsen and Scheraga 1975)

2n	N	Examples
1	1	
2	3	Tendamistate
3	15	BPTI
4	105	RNase A, Lysozyme
5	945	
6	10395	
7	135135	
.		
.		
.		
17	6332859870762850625	tPA, BSA, IgG

structure is caused by its mode of action. Kepler, in pondering about the six-cornered snow flake and its analogs in the biological world, assumed the code for structure formation to be laid down in Archimedes' regular bodies, apart from the "soul" of living things and other conceptions of the medieval mystical tradition.

In connection with the molecular description of structure formation, it has been obvious for many years that the self-organization of biomolecules is a spontaneous process which occurs autonomously, in analogy to physical processes such as the crystallization of ice or the formation of micelles of lipids in water. In the case of proteins, the one-to-one relationship of a one-dimensional structural information encoded in the amino-acid sequence along a given polypeptide chain, and the corresponding unique three-dimensional structure became obvious from Anson's observation that "hemoglobin which has been denatured in a variety of ways can be converted back into native protein" (Neurath et al. 1944; Anson 1945). Anfinsen, in following up Anson's work, extended this conclusion by reducing and reoxidizing the disulfide bridges in ribonuclease which were previously considered to determine not only the stability but also the spatial structure of the protein itself (Anfinsen 1966). A general conclusion to be drawn from the foregoing experiments has been: the three-dimensional structure of a protein is fully determined by its amino-acid sequence and the solvent environment; it represents the state of minimum potential energy (Anfinsen and Scheraga 1975).

Recent attempts to apply Anfinsen's approach to recover inactive protein deposited as "inclusion bodies" in overexpressing strains of bacteria included tissue plasminogen activator with 17 cystine bridges. As indicated in Table 1, the reoxidation and reorganization of the native structure (after preceding reduction in the randomized state) leads to just one out of 6332859870762850625 possible combinations of cysteine residues: available criteria prove the product of reconstitution to be indistinguishable from the initial native molecule (Table 2).

This result gives one representative example out of a great variety of different proteins where Anson's early experience has been established again and again. It corroborates the one-to-one relationship

Table 2. Comparison of native tissue plasminogen activator (tPA) with recombinant tPA from *E. coli* after reoxidation in the presence of glutathion[a]

	tPA from hamster cell culture	tPA from *E. coli* renatured
Molecular mass		
SDS PAGE	65,000[b]	58,000[c]
Sedimentation equilibrium	68,000	63,000
Stimulation by fibrinogen	20-30 fold	>20 fold
Antibody recognition	+	+
Spectral properties		
Fluorescence λ_{max} (native)	338 nm	338 nm
(denatured)	349 nm	349 nm
Stability range[d]		
Guanidine•HCl	≦3.3 ± 0.3 M	≦3.3 ± 0.3 M
Temperature	≦65°C	≦65°C

[a]Recombinant tPA deposited in the cells as inclusion bodies was solubilized in guanidine•HCl under reducing conditions and subsequently reoxidized in the presence of GSH/GSSG (Opitz 1988).
[b]Glycoprotein containing 7% carbohydrate.
[c]Carbohydrate-free
[d]Monitored by fluorescence emission at pH 7.5.

between amino acid sequence and three-dimensional structure. It does so in a way which is most illuminating for two reasons. (1) Obviously the disulfide bridges *stabilize* the native state rather than *determining* the spatial arrangement of the polypeptide backbone. (2) The domain structure of the antibody molecule which is known to be formed co-translationally (Bergman and Kuehl 1979) must be encoded in the amino acid sequence; there seems to be no requirement for either cellular components or vectorial processes determined by the directionality of translation. The latter conclusion has been confirmed by the total synthesis of ribonuclease using the Merrifield method. Synthesis in both directions, from the N- to the C-terminus (in vivo), and vice versa (in vitro) generates the native protein in its functional state (Gutte et al. 1972).

In summarizing these results we may conclude that folding in vivo and folding or reconstitution in vitro yield identical products. Thus, attempts to quantify folding and association under controlled conditions in vitro should be appropriate to elucidate the principles of protein self-organization.

Self-Organization of Proteins

In connection with in vitro folding and association the following questions need consideration: (1) In general, reconstitution is incomplete; recoveries obtained after careful optimization range from a few percent up to ≧90%. What are the side reactions competing with proper folding? (2) The time range observed in renaturation experiments may exceed the time range of cellular translation and structure formation by several orders of magnitude. For example the reshuffling of IgG takes up to 200 h while enzymes coming off the ribosome in *Escherichia coli* exhibit catalytic function within seconds. (3) In the case of multifunctional supermolecular structures the question arises whether there is an upper limit of size beyond which proteins require extrinsic factors or genetically determined assembly programs.

In the following we shall first deal with these questions considering the pathway of folding and association, as well as their kinetics. In subsequent paragraphs, a number of attempts to deduce topological patterns or rules for protein folding from known structures will be touched upon briefly. As will be shown, in searching for practical approaches allowing structure prediction in terms of a "code of protein folding", the structure-function relationship is indispensable, because so far no algorithm has been put forward which could be applied to "translate" a given amino acid sequence into its corresponding three-dimensional structure. "Knowledge-based" statistical approaches derived from topological *and* functional homologies of related proteins seem to be the only way to arrive at reliable structure *pre*dictions if one does not rest satisfied with *post*dicting known structures (Jaenicke 1987).

Folding and Reconstitution

The first two questions require a brief summary of the present state of our knowledge with respect to the mechanism of folding and association of proteins, and how this reflects the hierarchy of protein structure. The term protein folding refers to both the topology of the amino acids in the functional protein and the mechanism by which the unordered polypeptide chain achieves its native conformation in vivo as well as in vitro. Let us first consider the structural principles and then focus on mechanistic findings explaining the balance of folding and reconstitution.

Hierarchy of Protein Structure. The primary structure – the specific amino-acid sequence along the covalent polypeptide chain – is the basic level of organization (Fig. 1). The secondary and supersecondary structures define regular arrangements of the polypeptide backbone, e.g., α-helices and β-structures, and their energetically and topologically preferred aggregates in terms of bundles of helices, coiled coils, β-meanders, or twisted sheets, etc. The tertiary structure refers to the three-dimensional chain fold, with structural domains as spatially separated entities, sometimes with a specific function such as effector binding or intrinsic catalytic activity. The quaternary structure describes the stoichiometry and geometry of assemblies built up from subunits characterized by the foregoing levels of structural organization. Be-

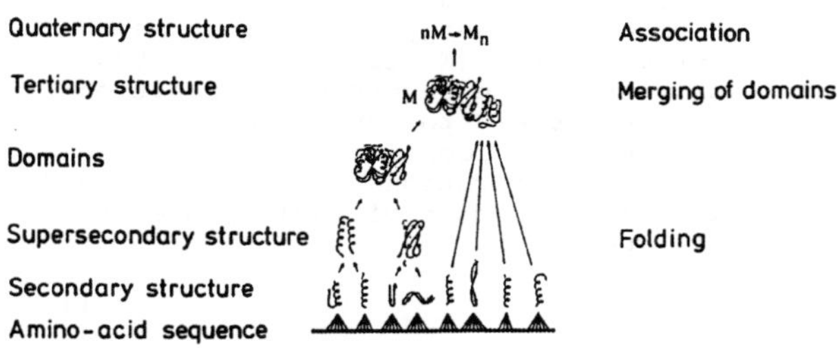

Fig. 1. Hierarchy of protein structure and protein folding

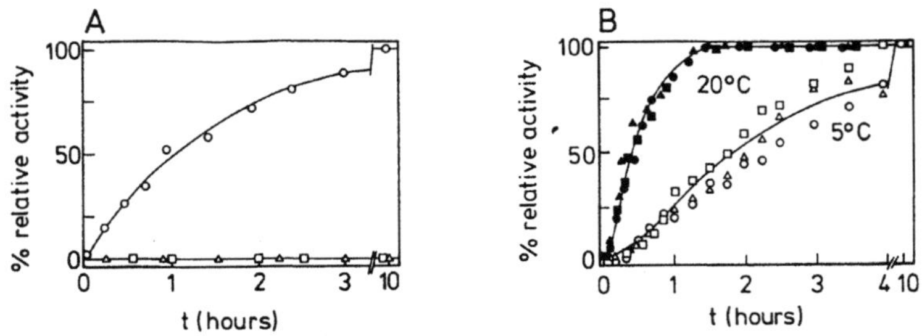

Fig. 2. (A) Reoxidation of reduced RNase A (o) and its fragments, PIR (Δ) lacking the C-terminal tetrapeptide, and V^- (□) lacking the C-terminus. 0.1 M Tris•HCl pH 8.2, 3 mM GSH, 0.3 mM GSSG, 1 mM EDTA. Enzyme concentration 3 μm, 25°C. (B) Reconstitution of the F34 (12-330) fragment of porcine muscle lactate dehydrogenase after denaturation in 6 M guanidine•HCl. 0.1 M phosphate buffer pH 7.6, 10 mM β-mercaptoethanol. Protein concentration: 1.4 (o,●), 4,5 (□,■), and 21 (Δ,▲) μg/ml. Enzyme activity measured at 25°C in the presence of 2 M $(NH_4)_2SO_4$. Profiles calculated according to $2M \xrightarrow{k_1} 2M' \xrightarrow{k_2} 2M \xrightarrow{\text{fast}} M_2$. (cf. Opitz et al. 1987)

yond a critical size of polypeptide chain, association represents a general structural principle extending the molecular description into the microscopic and macroscopic range.

In the given structural hierarchy, *folding*, as the transition from the one-dimensional to the three-dimensional structure, corresponds to the formation of the native secondary and tertiary structures, and *association* refers to the acquisition of the native quaternary structure. Both must be properly coordinated, because quaternary structure formation requires that the surfaces of the structural subunits be preformed in the correct way such that specific "recognition" is achieved.

Local or Global Minimum of Potential Energy? Whether folding occurs as a cotranslational process (and thus reflects the vectorial character of protein biosynthesis) cannot be answered in a general way (Fig. 2). Ribonuclease (RNase) may serve as an example where cleaving off the C-terminal tetrapeptide or even the terminal valine residue is sufficient to block the regain of activity after reduction/reoxidation (Teschner 1987); on the other hand, in a number of cases, the structure and function of enzymes have been shown to be indifferent to "circular permutation", as well as fragmentation (Creighton and Goldenberg 1984; Goldenberg and Creighton 1984; Kuchinke and Müller-Hill 1985; Fontana 1988). The question is essential because implicitly it favors the idea that proteins in their native state might occupy a local rather than the global minimum of potential energy, i.e., they may not necessarily be in their most stable state (Wetlaufer and Ristow 1973; Goldberg 1985). Previous analyses of the reversible transition between the native and denatured states seemed to contradict this view, since the transition curves were found to be identical regardless which probe was used to monitor denaturation/renaturation (cf. Jaenicke 1987, p. 134 f). Equilibrium data obtained for a number of small proteins clearly corroborated the "two-state model"

$$U \rightleftharpoons N, \qquad (1)$$

according to which the protein exists in one out of two states only, the polymorphous unfolded state (U), or the native state (N). Based on a variety of criteria, no partially or incorrectly folded intermediates should exist between N and U. Instead, the folding mechanism should consist of the random search for the thermodynamically most stable state.

Taking the given arguments to their logical conclusion (Levinthal 1968), one may ask how long it would take the unfolded polypeptide chain to test all possible conformations until, by chance, the correct native conformation, i.e., the global minimum of energy, would be reached. Considering only the polypeptide backbone with two rotatable bonds (ϕ,ψ) and three possible conformations about each of these bonds (120°, 240°, 360°), then RNase (124 amino acids) would allow 3^{246} conformations. Assuming a rate of bond rotation of 10^{13}/s, the time to sample each individual polypeptide conformation would be $t = 3^{246}/246 \times 10^{13} \sim 10^{102}$ s, or more than 3×10^{94} years. Since folding must occur during the lifetime of an organism, it can hardly occur by the given random search process. Correcting the oversimplyfied calculation for excluded volume effects (Dill 1985), the time needed for folding still remains much too long.

A way out of the dilemma has been mentioned: it is the view that the conformation of the final product of folding is determined by the *kinetic pathway* of folding. Based on this hypothesis, folding occurs by the fastest route available; it includes well-populated intermediates and generates the native state as the kinetically accessible state of minimum potential energy. This does not necessarily represent the global minimum belonging to the most stable state. If the two hypothetical conformers exist as (meta-)stable states, they should differ significantly. Otherwise it should be possible to convert one into the other within reasonable time. Since there is no way to analyze the native conformation in terms of the global energy minimum, the given arguments do not unambiguously prove a unique sequential pathway of folding.

Folding Pathways. The solution to the two-state or multi-state problem can only be expected from the kinetic analysis of the folding process which, for a number of single domain proteins, has clearly established the sequential scheme:

$$U \rightarrow I_i \rightarrow N \qquad (2)$$

with I_i as intermediates along the folding path. In the case of ribonuclease, detailed studies on the kinetics have shown that "the unfolded state" can be divided into two "substates": a fast-folding substate in which every proline residue is in the same cis or trans configuration as in the native protein, and a slow-folding substate in which at least one proline is the wrong isomer.

Analyzing the sequential folding pathway by spectral techniques, two intermediates have been identified. They are characterized by rapid local folding such that native secondary structure is recovered well before the protein acquires its native configuration (Schmid 1983). These stretches of secondary structure are unstable when exposed to the polar solvent.

However, by collapsing into a globular state (roughly approximating the native conformation), the polypeptide chain would eventually shield, and thus stabilize, its secondary structure. The collapsed state has been called the "molten globule" (Dolgikh et al. 1985; Ptitsyn 1987). It is characterized by a hydrodynamic volume higher than in the native molecule, allowing higher mobility of amino-acid side chains. Shuffling of the polypeptide chain by small movements within the molten globule then leads to the compact native state with minimum hydrophobic surface area.

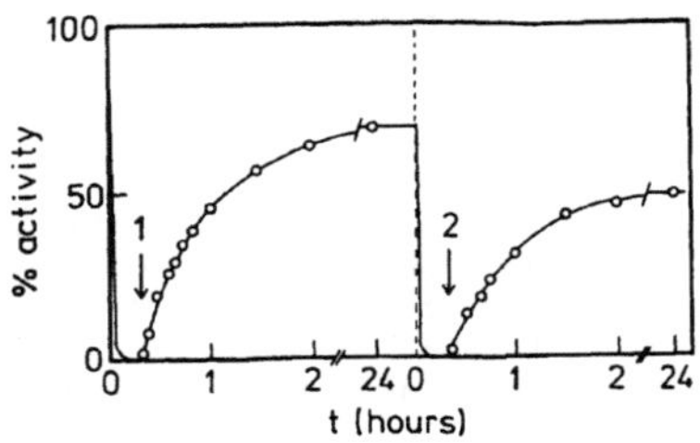

Fig. 3. Reactivation of octopine dehydrogenase from *Pecten jacobaeus* after repetitive denaturation in 6 M guanidine•HCl. 0.1 M phosphate buffer pH 7.6, 1 mM dithioerythritol, 1 mM EDTA. Enzyme concentration 0.9 µg/ml, 20°C. Rate constant for the slow folding step in the refolding scheme

$$\mathcal{U} \xrightarrow{\text{fast}} I \xrightarrow{k_1} N, \qquad k_1 = 5.4 \times 10^{-4}/s$$

In the case of multi-domain proteins, the merging of domains may lead to side reactions involving wrong domain interactions. To give an example, octopine dehydrogenase (M_r = 44,000, two domains) shows only 70% reactivation. A second denaturation/renaturation cycle again yields 70% (Fig. 3). The remaining 30% in each cycle represent inactive monomers with native-like secondary structure. Increased solvent viscosity has a drastic effect on reactivation, suggesting that domain merging is an essential step in the folding reaction (Teschner et al. 1987). The overall mechanism

$$\begin{array}{ccccccc} \mathcal{U} & \xrightarrow{\text{fast}} & U & \xrightarrow{\text{fast}} & I_N & \xrightarrow{\text{slow}} & N \\ & & \downarrow & & & & \\ & & U^* & & & & \end{array} \tag{3}$$

reflects the hierarchy of protein structure in that $\mathcal{U} \rightarrow U$ refers to the formation of secondary and supersecondary structure which subsequently interact to form "structured domains" ($U \rightarrow I_N$), or an inactive substate (U^*). Only I_N is able to condense further, generating the native state in a rate-determining (viscosity dependent) first-order process ($I_N \rightarrow N$). Most significant in the given context is the occurrence of two well-defined endproducts, U^* and N, which seem to prove that a unique amino-acid sequence may generate different stable conformations.

In proceeding from single chain molecules to oligomers, the complementation of the S-protein (amino acids 21-124) of RNase S with its S-peptide p (amino acids 1-20) may be used as a model reaction. As shown by kinetic experiments (including fast kinetics), concentration-dependent binding of p does not occur unless a structured intermediate has been formed:

$$U \rightleftharpoons I \underset{p}{\overset{p}{\rightleftharpoons}} pI \rightleftharpoons pN \text{ (RNase S)}. \tag{4}$$

The resulting consecutive reaction sequence, folding → association → activity, which can be fully described by one first-order and one second-order rate constant (k_1 = 10.2/s, k_2 = 7×10^5/M•s, represents the general mechanism observed for oligomeric and multimeric systems. It is obvious that in order to accomplish correct subunit association, folding has to provide specific intersubunit contacts. If collisions take place before "structured monomers" are formed, "wrong aggregation" ($U \rightarrow A$) occurs, thus diminishing the yield of quaternary structure formation according to the relative rates of the two competitive reactions:

$$A \leftarrow U \rightleftharpoons N. \tag{5}$$

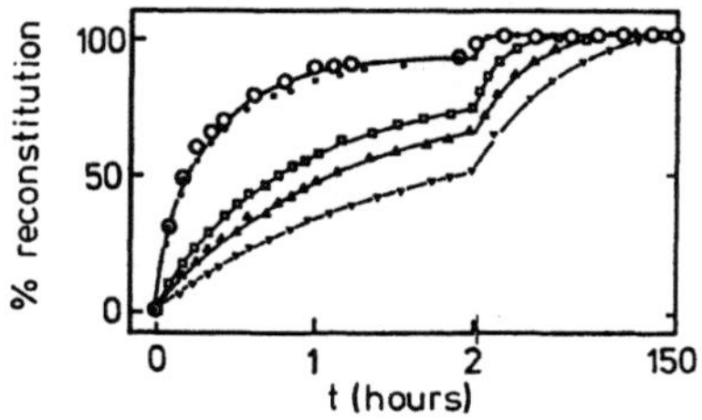

Fig. 4. Reactivation of lactate dehydrogenase from porcine heart muscle after deactivation at 1200 bar. 0.16 M Tris•HCl pH 7.6, 10 mM dithioerythritol, 1 mM EDTA, 20°C. Enzyme concentrations during reactivation (μg/ml): 10.8 (O), 6.0 (●), 1.4 (□), 1.0 (Δ), 6.0 (▼). Profiles calculated according to $4\,M' \xrightarrow{k_1} 4\,M \xrightarrow{fast} 2D \xrightarrow{k_2} T$ with $k_1 = 1.5 \times 10^{-3}$/s and $k_2 = 3.5 \times 10^3$/M•s. Recovery of native fluorescence and reactivation parallel each other

Obviously the competition of intra- and intermolecular processes can be eliminated by lowering the protein concentration, or by adsorbing the unfolded or nascent protein to a solid support, such as an ion-exchange resin or the ribosome.

At which state of the assembly process active species are formed is different for different systems; no rule can be given, even for closely related systems (cf. Jaenicke 1987). In all cases, the above-mentioned uni- bimolecular kinetic mechanism prevails, corresponding to the superposition of (unimolecular) folding with (bimolecular) association steps, according to

$$nM' \rightarrow nM \rightarrow \frac{n}{2} D' \rightarrow \frac{n}{2} D \rightarrow \frac{n}{4} T' \rightarrow \frac{n}{4} T \rightarrow\rightarrow\rightarrow, \qquad (6)$$

where n is the number of subunits, and M', M, D', D, T', T are monomers, dimers and tetramers in different conformational states (Fig. 4).

Folding Rates. Folding of small proteins both in vivo and in vitro takes place within seconds or less. Even large monomeric proteins, like Staphylococcal aldolase (M_r = 35,000), may acquire their native conformation surprisingly fast so that reactivation (after thermal denaturation) is complete within the mixing time of the enzymatic assay (Rudolph et al. 1985). However, this is the exception rather than the rule: in most cases shuffling of domains or subunits, especially when SH-SS exchange is involved, takes exceedingly long.

In trying to identify mechanisms which may accelerate the reaction, a number of alternatives come to mind:

1. Cotranslational, "vectorial" folding and in vivo domain formation may exclude or restrict side reactions on the folding path;

2. Specific ligands may serve to stabilize intermediates;

3. Enzymes may catalyze slow steps in protein folding: For example, peptidyl-prolyl-cis-trans-isomerase (PPI) has been shown to catalyze the cis-trans isomerization of X-proline peptide bonds (Fischer and Bang 1985; Lang et al. 1987), while protein-disulfide isomerase (PDI) catalyzes the thiol-disulfide exchange reaction (Freedman 1987). The suggestion that PDI is involved in proline isomerization has been refuted by simultaneous folding and reshuffling experiments: both enzymes act specifically and independently of one another (Fig. 5) (Lang and Schmid 1988).

4. Heat-shock proteins and related proteins (such as immunoglobulin heavy chain binding protein, BiP) found in normal unstressed cells from a wide variety of species seem to prevent aggregation or disrupt "wrong hydrophobic interactions", thus promoting folding and association (Pelham 1986).

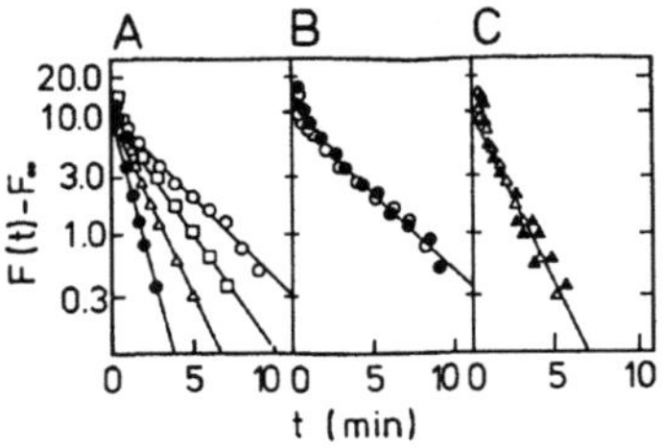

Fig. 5. Kinetics of slow refolding of the immunoglobulin light chain in the presence of varying concentrations of (A) prolyl isomerase (PPI), (B) disulfide isomerase (PDI) and (C) a mixture of both enzymes. Denaturation in 5 M urea, 0.1 M Tris•HCl pH 8.0, 25°C (disulfide bonds intact). Renaturation monitored by the decrease in fluorescence at 350 nm (λ_{exc} = 280 nm): 0.1 M Tris•HCl pH 7.8, 10°C. (A) PPI concentration: 0 (o), 0.07 (□), 0.21 (Δ), 0.34 (•) μM. (B) PDI concentration: 0 (o), 0.009 and 0.018 absorbance units at 280 nm (•). (C) Refolding in the presence of 0.21 μM PPI alone (Δ), and with 0.0034 and 0.017 absorbance units at 280 nm of PDI (▲). (For details see Lang et al. 1987; Lang and Schmid 1988)

Supermolecular Structures. In connection with the third question raised in the beginning, in vitro reconstitution of giant multienzyme complexes of phages and ribosomes, a variety of assembly mechanisms have evolved.

The observation that "joint reconstitution" of highly complex multienzyme complexes like pyruvate dehydrogenase ($M_r \sim 10^7$) generates the native assembly product shows that even a four-component system made up of about 60 subunits folds autonomously. Neither external factors nor a specific reaction sequence are required, apart from the inherent assembly program determined by the individual rate constants of the folding reactions of each of the four protein species involved in the formation of the heterologous multienzyme complex: intrinsic form determination governs both the assembly process and its highly symmetrical endproduct (Jaenicke and Perham 1982). This finding cannot be generalized: in conjugated systems of still higher complexity, or in complex polymorphic systems, like certain phages and viruses "aided assemblies" have been found which require either morphopoietic gene products or a specific reaction sequence in terms of "assembly maps" or an "assembly gradient". In the case of the ribosomal assembly this is provided by the growing RNA chain (Kellenberger 1984; Nomura and Held 1974; Nierhaus 1982). Form determination at the level of cellular organelles etc. is based on protein-protein and protein-nucleic acid interactions.

In order to provide conservation of structure, a well-defined arrangement of weak intermolecular interactions within a certain interacting area will account for specificity. Polymorphism is caused by conformational changes at the level of the tertiary structure of the subunits. Within complex systems containing various protein species, like phages or ribosomes, a whole hierarchy of interactions of varying bond strength is observed. Interactions decreasing in a specific sequence may be correlated with a corresponding sequence of assembly steps.

In considering the protein *folding* problem, evidently the "code" can only deal with systems governed by intrinsic form determination. Conjugated proteins as well as assembly systems involving extrinsic factors or the organization of the genome go beyond the scope of the present discussion.

Requirements for Folding. Cell-free in vitro translation has shown that correct folding is an intrinsic property of the nascent polypeptide chain. As mentioned, it occurs as a spontaneous and autonomous reaction requiring neither the intact cellular organization nor the native N-terminal initiation sequence. Cyclization of BPTI, altering connectivities in TSase and fragment studies on thermolysin, all three seem to indicate that folding starts as a highly cooperative process simultaneously in many segments along the polypeptide chain.

Structural Elements, Homology, and Divergence

Cooperative Units. "Seeds" for protein folding seem to consist of short stretches of α-helices and β-bends (Baldwin 1986), as well as loops (Fetrow et al. 1988). Since they involve only few residues, they are unstable and may appear and disappear rapidly. Evidence that sequences as short as nine amino acids are able to form α-helices and β-turns came from high-resolution NMR and circular dichroism spectroscopy. Using the C-peptide (residues 3-13) of RNase A, Brown and Klee (1971) were the first to show that there is significant helix formation at 0°C, while at ∿30°C randomization occurs. These and similar results by other authors seem to contradict predictions obtained from the Zimm-Bragg theory of α-helix formation which yields a minimum cooperative length of ∿20 amino-acid residues, regardless of the temperature and amino-acid composition. Obviously, specific interactions between side chains make an important contribution to the stability of isolated α-helices in their aqueous environment.

In fragment studies, there is always the question whether the isolated fragment is isomorphous within and without the intact protein, and furthermore, whether a given primary structure altogether yields a defined stable tertiary structure. From the examples given we may conclude that the structure of certain cooperative units persist when they are removed from their parent tertiary structure.

Structural Homology and Divergence. In defining relatedness of structural elements, a number of striking observation have been reported which clearly indicate that the protein folding code – if it exists – must be highly degenerate, i.e., a given tertiary structure may tolerate a great number of amino-acid exchanges without significantly changing its topology. To give a few examples, the all-α topology of the hemoglobins is preserved over the whole evolution of the globin family, although only two out of ∿150 amino-acid residues are found to be preserved. A similar result has been reported for the cytochromes where the homology amounts to less than 16%. In the case of homologous RNases with a wide variety of amino-acid exchanges (including variations in the number of proline residues), the spatial structure, as well as the pathway and the kinetics of folding, are found to be preserved (Krebs et al. 1983, 1985; Lang et al. 1986). Obviously, the folding mechanism is one of the crucial properties which has been optimized during evolution.

There have been numerous cases where sequences with very little or no detectable homology were found to adopt closely similar folds. To stress this point, it is illustrative to compare the estimated number of different proteins which is of the oder of 10^{12}, with the only four major categories of protein topologies deduced from X-ray data of known globular proteins (Richardson 1981).

In contrast to the observed degeneracy, identical segments within various unrelated sequences may have completely different conformations. To give an example, the sequence VNTFN has been shown to form an α-helix in erythrocruorin, whereas in RNase A it is part of a β-strand (Kabsch and Sander 1984). As shown by Wilson et al. (1985), there are numereous oligopeptides which exhibit this kind of structural ambiguity.

Short- and Long-Range Interactions

What becomes obvious from these twin structures, is one of the fundamental problems in protein structure prediction: both short- and long-range interactions are essential in defining the potential energy minimum. In this connection, it is the low "structural resolution power" of short-range interactions, and the importance of one-dimensional long-range interactions which cause ambiguities. It often occurs that the same region of a polypeptide chain is predicted with high probability both as α-helix and β-strand. This does not necessarily indicate incorrect parameters in the predictive methods, but it may mean that the short-range interactions alone actually do not determine the local conformation. In such a case, the ambiguity is removed in the actual folding process by the proper intervention of the long-range interactions. In determining the biologically active structure of a given amino-acid sequence, the dilemma of short-range vs. long-range interactions is crucial because a result of 90% correct prediction, or a 3 Å r.m.s. deviation, is satisfactory from the theoretical point of view; however, to predict the functional state one has to be >99% correct (Wetlaufer and Ristow 1973).

Stabilizing Forces

As has been mentioned previously, protein folding follows a unique sequential pathway where short stretches of secondary structure combine to compact local subdomains which collapse to form a "molten globule" intermediate, and finally the native tertiary and quaternary structure. The driving force of the overall process is the minimization of the free energy of stabilization

$$\Delta G_{stab} = \Delta H_b - T\Delta S_{solv} - T\Delta S_{chain}, \qquad (7)$$

where the sum of all intermolecular interactions (ΔH_b) and the entropy of solvation (in the aqueous solvent) have to compensate the decrease in chain entropy so that the polypeptide chain is arrested in its highly ordered native configuration. Considering the magnitude of the single terms in Eq. (7) ΔG_{stab} turns out to be a small difference of big numbers: values commonly determined for globular proteins are of the order of ∿40 kJ/Mol, so that proteins in their native state show only marginal stability. The finding that the contributions of intermolecular and intramolecular interactions are balanced in such a delicate manner explains that the ability of a given amino-acid sequence to form a stable three-dimensional structure is by no means trivial: out of the astronomically high number of possible sequences (20^n, with n = number of amino acids in the polypeptide chain) only a very small selection seems to be able to form a stable structure.

Summarizing the energetic contributions to protein stability, hydrogen bonds are crucial in determining the secondary and tertiary structure, although each protein hydrogen bond (due to the competition with the solvated unfolded chain) probably makes only a small contribution to the overall stability. However, because of their great number and because they tend to form cooperatively, their effect is considered to be significant.

The importance of electrostatic interactions between ions and/or dipoles in globular proteins is determined by the values to be assumed for the effective charges and dipole moments of chemical groups or structural elements, as well as the dielectric constant in the interior or at the surface of the protein. Ion pairs exert a stabilizing effect mainly on the tertiary and quaternary structure due to charge patterns at the protein surface. In the case of amide and helix dipoles it is obvious that the antiparallel alignment of helices and β-sheets is favorable in terms of dipole compensation.

Similarly, negative ions (such as phosphate) near the (positive) N-terminal end of helices have been found to have a stabilizing effect on a variety of proteins (Baldwin and Eisenberg 1987). The effect of glu^- ... lys^+ ion pairs on the secondary structure of proteins has been studied in detail using oligopeptides, 16 or 17 amino acids long (Marqusee and Baldwin 1987). Spacing the charged residues four or one residue apart

(i+4)E,K $AcAE^-AAAK^+E^-AAAK^+E^-AAAK^+ANH_2$

(i+4)K,E $AcAK^+AAAE^-K^+AAAE^-K^+AAAE^-ANH_2$,

significant helix stabilization is observed over the whole pH range (pH 2 - 12); reversal of the positions of glu^- and lys^+ confirms the stabilizing effect of ion dipol interactions: the ellipticity of the E,K peptides exceeds the helicity of the K,E peptides significantly, at least at low ionic strength.

The most important contribution to protein stability comes from hydrophobic interactions, i.e., the avoidance of water by the apolar side chains of globular proteins in aqueous solution. At low temperatures it is an entropy-driven phenomenon which may be adequately described by the atomic properties of the individual "hydrophobic residues" (Richards 1977; Rose et al. 1985a; Baldwin 1986). The outcome of estimates of the contribution of any atom of the protein to the free energy of stabilization based on solvent-accessible surface areas for folded proteins in comparison with unfolded or incorrectly folded polypeptide chains is a decrease in ΔG_{stab} of $\sim$4 kJ/mol of residue (Eisenberg and McLachlan 1986).

The only covalent bonds contributing to protein stability are disulfide bridges between cysteine residues. As mentioned previously, they are formed cotranslationally and may stabilize protein conformations which would otherwise be unstable. Attempts to improve protein stability by incorporating new disulfide bonds generally lead to stabilizing effects. However, due to the severe energetic constraints on disulfide bond geometry and unfavorable local effects of incorporating cysteine residues by mutation, only marginal stabilization has been achieved so far (Wetzel 1987; Creighton 1988). In certain cases, disulfides were not formed even though the proteins were folded and the thiols were adjacent. Obviously, the energy of cysteine formation depends on the redox potential of the (local) environment so that under physiological redox conditions reduced ribonuclease forms its native disulfide bridges, while "SH enzymes" such as lactate or malate dehydrogenase are kept in their reduced (active) state (Rudolph and Fuchs 1983).

Increments of all the given interactions (ΔH_b) contribute to protein stability, summing up to compensate the unfavorable conformational entropy (ΔS_{chain}). In cases where folding or association involve hydrophobic interactions, water release may lead to a significant increase in solvent entropy (ΔS_{solv}), thus enhancing the negative free energy of stabilization.

Structure Predictions

From the foregoing considerations we may predict that only a small fraction of all possible polypeptides with random amino-acid sequences are able to adapt folded conformations. Alterations either in the solvent (addition of denaturants) or in the sequence (site-directed mutagenesis) may easily alter the terms in Eq.(7) such that ΔG_{stab} alters its sign.

As mentioned, the free energy of the protein in its folded state is determined by all the physical interactions within and without the protein, plus all entropic terms. The weak interactions involved in stabilizing the native state cooperate simultaneously. They contribute to the extent to which they are more favorable than the corresponding interactions between the unfolded state and the solvent. In this connection, compacting the polypeptide chain to a state of mimimum solvent accessible surface area is of utmost importance in stabilizing the folded state.

It is obvious that the subtle balance of intra- and intermolecular interactions contributing to ΔG_{stab} poses an unsurmountable problem to any attempts to predict stable protein structures ab initio, i.e., starting from mere amino-acid sequences. It is trivial that protein molecules are too large to be treated by quantum and statistical mechanics. Due to the marginal energies of stabilization compared to the total internal energy (ratio $\sim$ 1:10^6 for small proteins), the quantum mechanical energy would have to be computed to an accuracy of $\sim$1ppm in order to make useful predictions. However, even this would not necessarily lead to a meaningful result, because (1) the solvent environment and its entropic contributions may be decisive, and (2) the native state may be controlled by thermodynamic *and* kinetic constraints. As one would predict, the perspective becomes brighter if we consider local conformations within a structurally invariant frame work. To give an example, complementarity-determining regions in the antibody binding site can be predicted with a high degree of precision (Rees 1987).

Statistical Approaches

Statistical analyses of known structures have been successfully applied to derive quantitative information about the conformational and positional properties of amino-acid residues in globular proteins. This has provided insight into the rules which govern the spatial organization. Determining the probabilities with which the different amino acids occur in α-helices, β-strands and "nonregular structures" on one hand, and at the surface or in the interior of the protein, on the other, conformational or positional "phase diagrams" have been devised (Levitt 1978; Pohl 1980). The outcome of this kind of analysis has been: (1) All amino acids (with arginine as the only exception) show a clear preference for only one secondary structure; no amino acid favors two types. (2) The ratio of the internal and external fraction describes the distribution of the different amino-acid residues between the hydrophobic core and the aqueous phase at the surface of a given protein. It shows good correlation with the thermodynamic transfer data describing the relative solubilities of amino acids in water and less polar solvents determined by Nozaki and Tanford (1971). Detailed normalized distributions of accessibility to solvent for all 20 amino-acid residues calculated from proteins of known structure are illustrated in Fig.6. The data show clearly that in the folding process there are roughly equivalent decreases in the accessibility of both the polar and nonpolar groups.

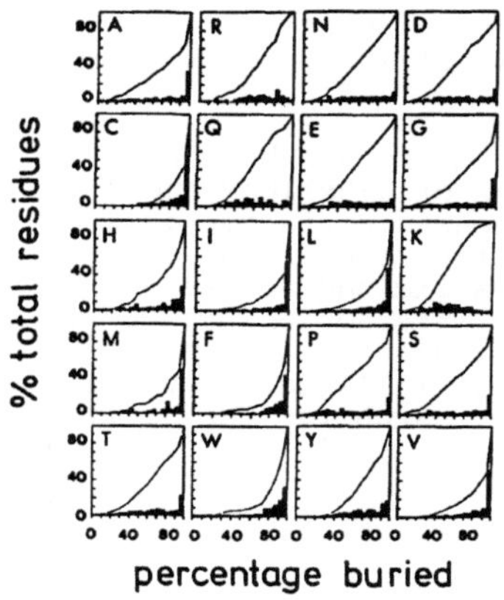

Fig. 6. Normalized distribution functions of accessibility to solvent for the 20 residues of amino acids calculated from proteins of known structure. Each function is a histogram showing the percentage of residues of that type that are: full accessible, 5% buried, 10% buried ... completely buried. An integral curve is superimposed on the histograms. (Rose et al. 1985a)

There exist numerous approaches to predict elements of secondary structure from known amino-acid sequences. Since the early test case of adenylate kinase where Schulz et al. (1974) for the first time reported true predictions, algorithms have been improved considerably so that the accuracy of helix prediction now reaches 65-70% (for review see Schulz 1988). In general, α-helix predictions are superior over predictions of β-structures for obvious reasons: the α-helix is fully determined by next-neighbor interactions, while β-sheets are hair-pin-like structures the formation of which requires well-defined turns, so that the prediction involves distant segments of the polypeptide chain, as well as intermolecular interactions of different chains.

Next to α-helices and β-strands, reverse turns and Ω loops are most abundant in globular proteins (Richardson 1981; Rose et al. 1985b; Fetrow et al. 1988). Since they contain mostly hydrophilic residues they are normally located at the surface of proteins. Although they are most important in accomplishing a compact globular structure, they seem to play only a passive role in the folding process. Energetically they are points of least resistance since they form maximally one hydrogen bond. Although they are usually easy to localize, detailed structure predictions are most difficult.

Packing

The next-higher level of structure prediction refers to the packing or "docking" of the predicted α-helices and β-strands into supersecondary structures and a native-like tertiary fold (Richards 1977; Sternberg 1983; Scheraga 1983). A variety of docking procedures have been proposed (Wodak et al. 1987). In general, empirical rules for the packing of secondary structural elements were obtained from known structures; in addition topological and energetic considerations were used to select a set of possible structures which were then applied for further refinement, for example using simplified energy calculations. In this context, it is obvious that the packing efficiency of folded polypeptide chains may be improved by protein-protein interactions beyond the monomer level. More than in the previous discussion, theoretical calculations and computer simulations are hampered by the large size of the molecules. Restricting the number of molecular states to structures close to the free components (subunits), visual inspection of molecular models may be combined with conformational energy calculations, making use of interactive computer graphics methods. Detailed surveys of such procedures have been summarized by Wodak et al. (1987).

Optimal sequential stability due to intra- and interchain hydrogen bonds provides clear preference to the formation of α-helices and β-sheets. The association of these structural elements to larger assemblies may involve a decrease in hydrophobic surface area. Applying these criteria to a given amino-acid sequence, a limited number of supersecondary structures involving helix-helix interactions and β-structures may be predicted which allows us to reduce the number of "allowed" structures drastically (<100 out of $\sim 10^8$) (Richards et al. 1980; Cohen et al. 1980). Further refinement to search for the native-like tertiary fold may then be accomplished by energy minimization and molecular dynamics simulation methods.

Energy Minimization and Molecular Dynamics

To study the energetics, the protein potential energy surface, i.e., the energy of the molecule as a function of the atomic coordinates, is required. This can then be used to determine the relative stabilities of different possible structures of the protein. Since the potential energy surface of a protein molecule is not accessible from quantum mechanical calculations, empirical energy functions have to be used. In this context, methods developed to calculate the structure of small organic molecules may also be applied to a protein. However, there is one important difference which has already been mentioned in connection with the stabilizing forces: in proteins (in contrast to small molecules) contacts between nonbonded atoms play an essential role in the potential energy of the folded structure.

The forces between single atoms may be obtained from the first derivatives of the potential energy with respect to the atomic positions; they may be calibrated according to known properties of small molecules. Given an empirical potential energy function, energy minimization allows us to search for energy minima determining the most stable configurations of a molecule. They do not necessarily include the real native structure, because this may simply not be among the lowest potential energy states, for two reasons: (1) certain conformations may be entropically favored and therefore have a lower energy, and (2) the "global" free energy minimum may not be accessible on the pathway of folding. For macromolecular systems, the vast number of local minima does not allow the exhaustive search for the whole energy surface to discover the global energy minimum. Therefore, a local minimum close to the (known) X-ray structure is examined. If the chosen energy function does not provide the correct structure of the standard protein (for example BPTI), it will be modified until the conformation with minimum potential energy and the native structure coincide.

This "Darwinian approach" in developing the optimum energy function is prone to a certain ambiguity as soon as one goes from structure simulation to structure prediction. In the given context, a considerable improvement may come from the molecular dynamics approach which allows both the configurational entropy and the solvent to be included into the calculation.

In molecular dynamics we focus our attention on the empirical fact that a "native protein" at room temperature (due to its internal motions) samples a range of conformations in the general neighborhood of the native average structure. As shown for the accessibility of the heme-pocket in myoglobin and hemoglobin for oxygen and other protein-ligand interactions, structural fluctuations about the average structure are most significant in protein function.

In molecular dynamics simulations, the energy function applied in energy minimization calculations is used to obtain the forces acting on the protein atoms by taking the appropriate derivatives using Cartesian coordinates. They allow us to calculate the respective positions and accelerations of all the atoms in the system at any time. Integrating the forces acting on all atoms at various time intervals ($\Delta t \sim 1$ ps), a trajectory is generated which describes the internal motions of all the atoms as a function of time.

Applying both energy minimization and molecular dynamics to basic pancreatic trypsin inhibitor (BPTI) as an example, energy minimization converges to a precise structure with a well-defined energy value. The calculated equilibrium conformations are found to be sensitive to the chosen values of the energy parameters so that the native X-ray conformation can be reproduced (Levitt 1983). Molecular dynamics simulation proves that the mean dynamic conformation is as close to the experimental three-dimensional structure as the static equilibrium conformation. Obviously, in the case of BPTI, the entropic effects included in dynamic simulations have only an insignificant effect on the calculated conformation of the protein.

Considering the subtle interplay of intra- and intermolecular interactions in native proteins, one would expect that it is essential to take the aqueous solvent into consideration in any structure simulation. Generally, it is possible to include the solvent molecules into molecular dynamics calculations, either explicitely as well-specified hydration water molecules, or in terms of a "stochastic boundary". In applying such approaches, the atomic fluctuations of the protein atoms relative to their average positions were found to be in agreement with temperature factors determined by X-ray crystallography. One would expect that results referring to the protein in vacuo (ignoring the solvent) are meaningless. However, when experimental and calculated fluctuations from energy minimization and molecular dynamics simulations are compared, one is surprised to find qualitative agreement. Prominent discrepancies are only observed for charged side-chains on the surface of the molecule. In the interior, environmental effects on the amplitude of fluctuations are found to be small. Taking the solvent or crystal field into account, the average structure comes significantly closer to the X-ray structure compared with the result in vacuo: the r.m.s. deviation decreases from >2 Å to $\sim$1 Å (van Gunsteren et al. 1983).

Knowledge-Based, Computer-Aided Structure Predictions

In summarizing all available methods to predict the three-dimensional structure of proteins from the amino-acid sequence, it becomes clear that presently semi-empirical combinatorial approaches are the only way to arrive at a satisfactory result. According to the previous discussion, a hierarchic progress from "joint predictions" of structural motifs, via "template search" to simplified, and finally refined energy calculations seems to be appropriate. A most useful tool at all stages of the given procedures has been interactive three-dimensional computer graphics methods. Their data base has been obtained mainly from X-ray crystallography and sequence homology or some general analogy with known crystal structures making use of the basic motifs of protein architecture (Blundell and Sternberg 1985).

The evolutionary mechanism of gene duplication and divergent evolution leads to families of homologous proteins that share similar primary and three-dimensional structures. This provides a means of predicting the conformation of one protein based on the crystal coordinates of a homologous protein. The process starts with the alignment of the two sequences during which insertions and deletions are generally placed in the loop regions connecting the regular secondary structural elements, the α-helices and β-sheets, as these structures tend to be more conserved during evolution. This is followed by model building (interactive computer graphics) in which the known crystal structure is initially taken and alterations in the amino-acid sequence are subsequently considered step by step. This starting model is then adjusted to yield a stereochemically sensible conformation in which (1) there are no forbidden close contacts, (2) there is tight packing of the hydrophobic core, and (3) polar and charged atoms that do not form internal hydrogen bonds or salt bridges are exposed to the solvent.

In the case of proteins for which there is no evidence of sequence homology, the limited number of structural motifs characterized by Richardson (1981) may be used to recognize specific key sequences dictating chain folding into one or the other pattern of tertiary structure. In this context, secondary structure predictions and empirical rules governing the packing of α-helices and β-structures may be combined with computer-aided approaches.

Effects on Folding Not Encoded in the Amino-Acid Sequence

In trying to gain some deeper understanding of the reasons why structure predictions for proteins commonly do not yield an unambiguous result, the question has been raised whether the native structure of a protein molecule is determined exclusively by its primary structure (including the intrinsic folding pathway). Three types of experiment seem to confirm the hypothesis: (1) in vitro reconstitution experiments after prior denaturation, (2) solid-phase peptide synthesis (from the C- to the N-terminus), and (3) semi-synthetic linking of protein fragments. However, there are at least two recent observations which indicate that approaches based on the amino-acid sequence alone may not be sufficient because other components may modulate the folding process.

Codon Usage

Considering the amino-acid pools and the codon usage in a given organism, peculiarities of the nucleotide sequence of the messenger RNA may affect the folding of the nascent polypeptide chain by influencing the rate of protein biosynthesis. The rates of in vivo translation and folding, and the way both processes are tuned within the cell are not known in detail. The question as to which process is rate-determining under specific conditions must be of importance in connection with the efficiency of folding. Evidence that there might be discontinuities in the rate of translation has recently been presented (Varenne and Lazdunski 1986). It is suggestive to assume that translational pauses in peptide segments connecting domains may be a device allowing to gear how the polypeptide chain "folds by parts" (Wetlaufer 1981). If this mechanism turns out to be of general importance, it is clear that codon usage must have significant implications for accurate folding of recombinant proteins in heterologous systems, since the codon bias of the original organism is unlikely to be reflected in the tRNA pools of the host.

Effects of Ligands

A completely different mechanism which again may direct and regulate the folding of the nascent polypeptide chain (as it directs reconsti-

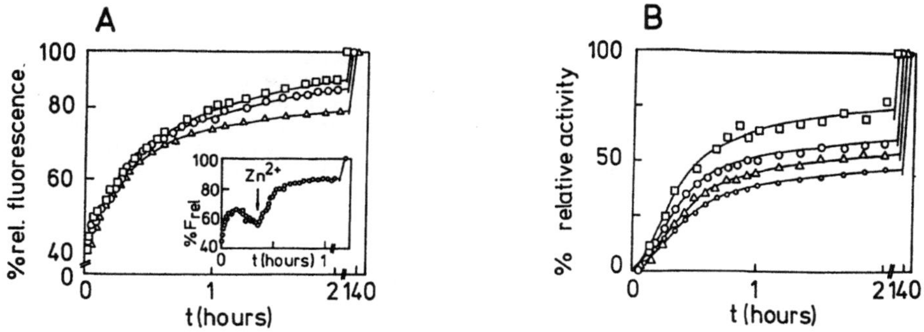

Fig. 7. Reconstitution of horse-liver alcohol dehydrogenase after 15 min denaturation in 6 M guanidine•HCl, 0.1 M phosphate buffer pH 7.6, 0.5 mM dithioerythritol, 20°C. (A) Regain of native fluorescence (λ_{exc} = 275 nm, λ_{em} = 330 nm) in the same buffer; enzyme concentrations during reconstitution (µg/ml): 8.6 (Δ), 12.5 (O), 19.0 (□). *Insert* Effect of 10 mM Zn^{2+} (↓) on the renaturation of the denatured and metal-free enzyme. (Gerschitz et al. 1978). (B) Reactivation in the same buffer, plus 0.1 mM NAD^+, 25°C at varying protein concentration (µg/ml): 6.1 (o), 9.0 (Δ), 11.0 (O), 20.0 (□)

tution after previous unfolding) involves ligands. As in the case of codon usage, effects not inherent in the amino-acid sequence are found to determine the pathway as well as the product of folding.

Ligands are expected to affect the mechanism of folding mainly by stabilizing intermediates or the final product, thus shifting the equilibrium toward the native state. Accordingly, complex formation will lead to an improved yield of the product of folding or reconstitution. Considering coenzymes or substrates as potential primers, the rate of folding will be influenced to a lesser extent because high-affinity binding in these cases requires a native-like state, i.e., more or less the final product of folding. This holds, for example, for coenzyme binding to NAD- or PLP-dependent enzymes, where formation of the holoenzyme has a stabilizing rather than a structure-promoting effect. In the case of small ligands such as metal ions the situation is different. As shown for horse-liver alcohol dehydrogenase, the essential Zn^{2+} ions not only stabilize the native structure and participate in catalysis; they also provide "seeds for folding" and serve to fix folding intermediates (cf. Jaenicke 1987). The structural and functional aspects may be correlated with the "catalytic" and functional Zn^{2+} ions present in each of the two subunits of the enzyme. Folding in the absence of Zn^{2+} yields neither catalytic function nor proper assembly, because structured intermediates with native-like fluorescence properties are trapped by forming "wrong aggregates". The side reaction is quenched immediately after Zn^{2+} is added; at the same time, enzymatic activity is generated in a second-order reaction, thus proving that the metal ion plays an essential role in keeping the folding intermediates on the correct folding path (Fig.7).

Finally, other proteins may be considered as ligands involved in protein structure formation. A few examples may illustrate this: (1) Separate domains of lactate dehydrogenase obtained by limited proteolysis have been found to "recognize" each other so that catalytically active "nicked dimers" are obtained upon renaturation. Because of the mutual stabilization of the fragments, the reaction requires "joint reconstitution" of the two domains (Opitz et al. 1987). (2) In Nagai's studies on recombinant hemoglobins expressed in *Escherichia coli* (Nagai et al. 1985), high yields of functional tetramers were obtained by using one chain as a "folding matrix" for the other (K. Nagai personal communication). (3) In the case of pyruvate dehydrogenase from *Bacillus stearothermophilus*, both lipoate acyltransferase (E2) and lipoamide dehydrogenase (E3) exhibit independent folding kinetics; regaining the overall catalytic activity requires the presence of all four components: obviously the "core" serves to stabilize the multienzyme complex in its native state (Jaenicke and Perham 1982).

Conclusions

Faced with the astronomically high number of possible conformers of a polypeptide chain of average length, a surprisingly small number of "folding topologies" has been deduced from well over 100 known high-resolution crystal structures solved so far. Obviously a given tertiary structure may tolerate a great number of amino-acid exchanges without significantly changing its topology: the "code of protein folding" must be highly degenerate. On the other hand, single amino-acid substitutions can lead to large alterations in the stability and folding, indicating that certain parts of the polypeptide chain contribute important interactions to overall stability. Compared to the number of possible amino-acid sequences, only a small number of polypeptide chains codes for a defined functional protein.

In vitro folding experiments have shown that small single-domain proteins may undergo fully reversible denaturation reactions. In multidomain proteins, the acquisition of the native three-dimensional structure is determined by folding and merging of domains. To generate assembly structures, sequential folding and association must be coordinated such that specific recognition of "structured monomers" is accomplished.

Apart from the complex interplay of short-range and long-range interactions of the amino-acid residues along the polypeptide chain, structure predictions of proteins on the basis of their primary structure may be hampered by codon usage, amino-acid pools, and site-specific ligands. They are specific for any given organism and any given gene product, so that approaches based solely on the physical properties of amino acids may on principle not be sufficient to predict the three-dimensional structure of proteins.

Acknowledgments. Supported by grants of the Deutsche Forschungsgemeinschaft and the Fonds der Chemischen Industrie.

References

Anfinsen CB (1966) Harvey Lect 61:361-384
Anfinsen CB, Scheraga HA (1975) Adv Protein Chem 29:205-300
Anson ML (1945) Adv Protein Chem 2:361-384
Baldwin RL (1986) Trends Biochem Sci 11:6-9
Baldwin RL, Eisenberg D (1987) In: Oxender DL, Fox CF (eds) Protein engineering. A.R. Liss, New York, pp 127-148
Bergman LW, Kuehl WM (1979) J Supramol Struct 11:9-24
Blundell TL, Sternberg MSE (1985) Trends Biotechnol 3:228-235

Brown JE, Klee WA (1971) Biochemistry 10:470-476
Buchner J (1987) Diplomarbeit, University of Regensburg
Cohen FE, Sternberg MJE, Taylor W (1980) In: Jaenicke R (ed) Protein folding, Elsevier, North Holland, Amsterdam, pp 131-148
Creighton TE (1988) Bio Essays 8:57-63
Creighton TE, Goldenberg DP (1984) J Mol Biol 179:527-545
Dill KA (1985) Biochemistry 24:1501-1509
Dolgikh DA, Abaturov LV, Bolotina IA, Brazhnikov EV, Bushnev VN, Bychkova VE, Gilmanshin RI, Lebedev YO, Semisotnov GV, Tiktopulo EI, Ptitsyn OB (1985) Eur Biophys J 13:109-121
Eisenberg D, McLachlan AD (1986) Nature 319:199-203
Fetrow JS, Zehfuß MH, Rose GD (1988) Bio/Technology 6:167-171
Fischer G, Bang H (1985) Biochim Biophys Acta 828:39-42
Fontana A (1988) Biophys Chem 29:181-193
Freedman RB (1987) Nature 329:196-197
Gerschitz J, Rudolph R, Jaenicke R (1978) Eur J Biochem 87:591-599
Goldberg ME (1985) Trends Biochem Sci 10:388-391
Goldenberg DP, Creighton TE (1984) J Mol Biol 179:497-526
Gutte B, Lin MC, Caldi DG, Merrifield RB (1972) J Biol Chem 247:4763-4767
Jaenicke R (1987) Prog Biophys Mol Biol 49:117-237
Jaenicke R, Perham RN (1982) Biochemistry 21:3378-3385
Kabsch W, Sander C (1984) Proc Natl Acad Sci USA 81:1075-1078
Kellenberger E (1984) Helv Phys Acta 57:188-201
Kepler J (1611) Strena seu de nive sexangula. Tampach G, Francofurti ad Moenum
Krebs H, Schmid FX, Jaenicke R (1983) J Mol Biol 169:619-635
Krebs H, Schmid FX, Jaenicke R (1985) Biochemistry 24:3846-3852
Kuchinke E, Müller-Hill B (1985) EMBO J 4:1067-1073
Lang K, Schmid FX (1988) Nature 331:453-455
Lang K, Wrba A, Krebs H, Schmid FX, Beintema JJ (1986) FEBS Lett 204:135-139
Lang K, Schmid FX, Fischer G (1987) Nature 329:268-270
Levinthal C (1968) J Chim Phys 65:44-45
Levitt M (1978) Biochemistry 17:4277-4285
Levitt M (1983) J Mol Biol 170:723-764
Marqusee S, Baldwin RL (1987) Proc Natl Acad Sci USA 84:8898-8902
Nagai K, Perutz MF, Poyart C (1985) Proc Natl Acad Sci USA 82:7252-7255
Neurath H, Greenstein JP, Putnam F, Erickson JO (1944) Chem Rev 34:157-265
Nierhaus KH (1982) Curr Top Microbiol Immunol 97:82-155
Nomura M, Held W (1974) In: Nomura M, Tissière A, Lengyel P (eds) Ribosomes, Cold Spring Harbor Laboratory, pp 193-223
Nozaki Y, Tanford C (1971) J Biol Chem 246:2211-2217
Opitz U (1988) Dissertation, University of Regensburg
Opitz U, Rudolph R, Jaenicke R, Ericsson L, Neurath H (1987) Biochemistry 26: 1399-1406
Pelham HRB (1986) Cell 46:959-961
Pohl F (1980) In: Jaenicke R (ed) Protein folding. Elsevier, North Holland, Amsterdam, pp 183-197
Ptitsyn OB (1987) J Protein Chem 6:273-293
Rees AR (1987) Immunol Today 8:44-45
Richards FM (1977) Annu Rev Biophys Bioeng 6:151-176
Richards FM, Richmond TJ, Sternberg MJE, Cohen FE (1980) In: Jaenicke R (ed) Protein folding, Elsevier, North Holland, Amsterdam, pp 117-129
Richardson FS (1981) Adv Protein Chem 34:167-339
Richardson JS (1984) In: Wetlaufer DB (ed) The protein folding problem, Westview, Boulder, pp 1-28
Rose GD, Geselowitz AR, Lesser G-J, Lee RH, Zehfuß MH (1985a) Science 229:824-839
Rose GD, Gierasch LM, Smith JA (1985b) Adv Protein Chem 37:1-109
Rudolph R, Fuchs I (1983) Hoppe-Seyler's Z Physiol Chem 364:813-820
Rudolph R, Bohrer M, Fischer S (1985) Eur J Biochem 131:181-186
Scheraga HA (1983) In: Clementi E, Ramaswamy H, Sarma RH (eds) Structure and dynamics: nucleic acids and proteins. Adenine, New York, pp 463-476

Schmid FX (1983) Biochemistry 22:4690-4696
Schulz GE (1988) Annu Rev Biophys Chem 17:1-21
Schulz GE, Barry CD, Friedman J, Chou PY, Fasman GD, Finkelstein AV, Lim VI, Ptitsyn OB, Kabat EA, Wu TT, Levitt M, Robson B, Nagano K (1974) Nature 250:140-142
Sternberg MJE (1983) In: Geisow, Barrett (eds) Computing in biological science, Elsevier Biomed, Amsterdam, pp 143-177
Teschner W (1987) Dissertation, University of Regensburg
Teschner W, Rudolph R, Garel J-R (1987) Biochemistry 26:2791-2796
van Gunsteren WF, Berendsen HJC, Hermans J, Hol WGJ, Postma JPM (1983) Proc Natl Acad Sci USA 80:4315-4319
Varenne S, Lazdunski C (1986) J Theor Biol 120:99-110
Wetlaufer DB (1981) Adv Protein Chem 34:61-92
Wetlaufer DB, Ristow S (1973) Annu Rev Biochem 42:135-158
Wetzel R (1987) Trends Biochem Sci 12:478-482
Wilson IA, Haft DH, Getzoff ED, Tainer JA, Lerner RA, Brenner S (1985) Proc Natl Acad Sci USA 82:5255-5259
Wodak SJ, De Crombrugghe M, Janin J (1987) Prog Biophys Mol Biol 49:29-63

The Method of Protein Structure Determination by NMR in Solution: Initial New Insights Relating to Molecular Mobility

K. Wüthrich [1]

Introduction

Protein design and protein engineering are the central theme of this Mosbacher Colloquium. In the wake of the development of these new research areas we have witnessed a dramatic upsurge of the interest in studies of three-dimensional protein structures, which is also clearly manifested in the program. The driving force behind this can readily be appreciated from the fact that the number of sequence variations for a polypeptide chain with n amino acid residues is 20^n, which virtually rules out systematic screening as a strategy for the design of new or improved proteins. All the more emphasis is on approaches guided by basic knowledge on the relations between structure and functions of natural proteins. In this context it is quite timely that a second method for protein structure determination, besides the long-established diffraction techniques with single crystals, has recently become available, which makes use of data collection by nuclear magnetic resonance spectroscopy (NMR) in solution (Wüthrich 1986).

The introduction of NMR as a second approach to protein structure determination is of fundamental interest, since it provides data that are in many ways complementary to those obtained from X-ray crystallography. Its use thus promises to widen our view of protein molecules with regard to a better grasp of the relations between structure and function. The complementarity of the two methods results from the facts that the time scales of the two types of measurements are widely different (Blundell and Johnson 1976; Wüthrich 1986), and that in contrast to the need of single crystals for diffraction studies the NMR measurements use proteins in solution or other noncrystalline states.

The NMR Method for Protein Structure Determination

In present practice, the data for protein structure determination are collected using two-dimensional (2D) NMR experiments (Ernst et al. 1987; Wüthrich 1986) (Fig. 1). 2D NMR spectra are presented in a two-dimensional frequency plane, with the two frequency axes ω_1 and ω_2. The spectra of prime importance for work with proteins contain an array of *diagonal peaks* with $\omega_1 = \omega_2$, which correspond largely to the conventional one-dimensional (1D) spectrum and display the chemical shift positions of the resonance lines. In addition, there are a large number of *cross peaks* with $\omega_1 \neq \omega_2$. Through simple geometric patterns, as indicated in Fig. 1A for the peaks labeled i, j, and k, each cross-peak establishes a correlation between two diagonal peaks. In 2D nuclear Overhauser enhancement spectroscopy (NOESY), which is the pivotal

[1] Institut für Molekularbiologie und Biophysik, Eidgenössische Technische Hochschule-Hönggerberg, CH-8093 Zürich, Switzerland

39. Colloquium Mosbach 1988
Protein Structure and Protein Engineering

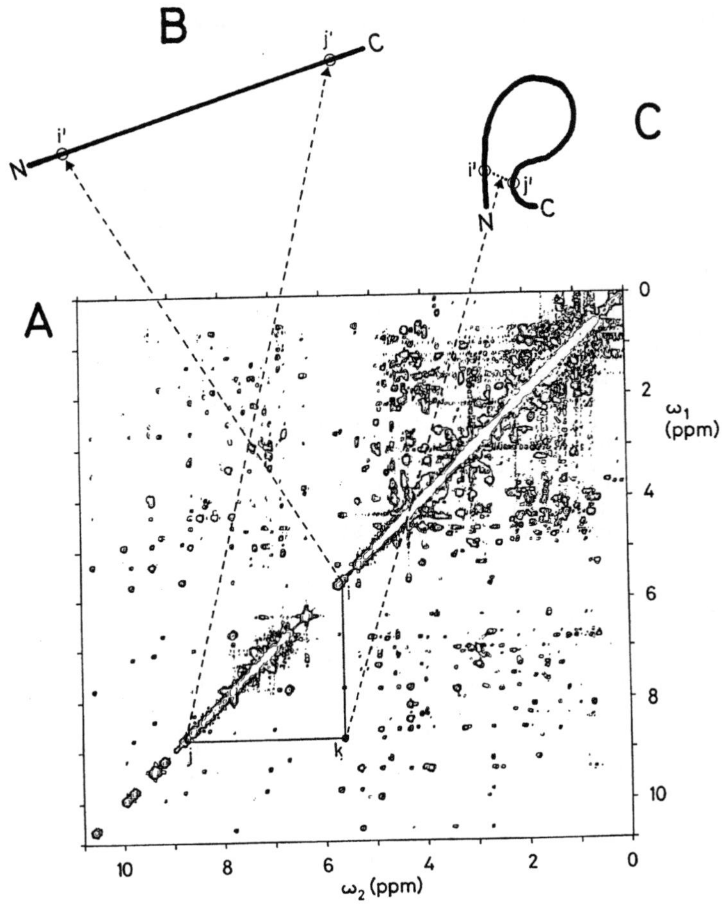

Fig. 1. Survey of the NMR method for protein structure determination in solution. (A) Contour plot of a 500 MHz ^{1}H NOESY spectrum of the protein basic pancreatic trypsin inhibitor (BPTI), with the two frequency axes ω_1 and ω_2. A cross-peak *k* is connected by a horizontal and a vertical line with the diagonal peaks *i* and *j*. (B) *Straight line* representing a polypeptide chain. Two protons are identified by *circles* and the letters *i'* and *j'*. *Two broken arrows* connect the corresponding diagonal peaks *i* and *j* in the NOESY spectrum with these protons. (C) Schematic presentation of the loop structure imposed on the polypeptide chain by the fact that the protons i' and j' are separated by a distance of less than 5.0 Å, as is evidenced by the appearance of the NOESY cross-peak k

experiment in studies of protein structures, the cross-peaks represent NOE's and indicate that the protons corresponding to the two correlated diagonal peaks are separated only by a short distance, say less than 5.0 Å. This information is used in the following way: First, sequence-specific resonance assignments must be obtained, i.e., for all the protons in the polypeptide chain the corresponding diagonal peaks must be identified. This is indicated in Fig. 1, A and B, by the two dashed arrows linking the protons i' and j' with their diagonal posi-

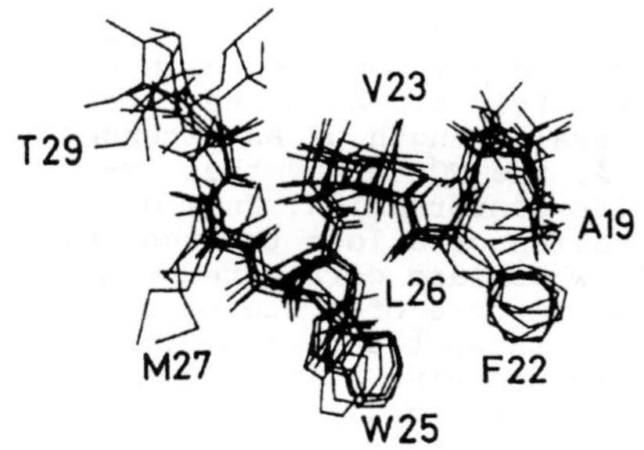

Fig. 2. Three-dimensional structure of the segment consisting of the residues 19-29 in the polypeptide hormone glucagon bound in the lipid-water interphase on the surface of dodecylphosphocholine micelles. Six conformers have been superimposed for minimal RMSD, each of which was computed with a metric matrix distance geometry algorithm from the same NMR data but with different starting conditions. The side chains of the residues Gln20, Asp21, Gln24 and Asn28 are not shown, since they were not constrained by the experimental data. (Braun et al. 1983)

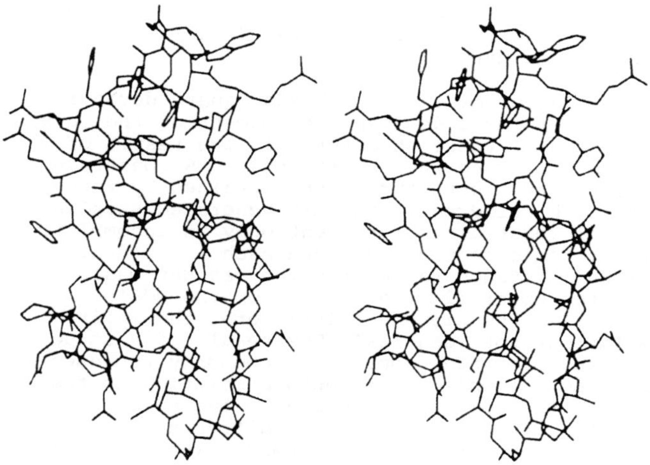

Fig. 3. Stereo view of the three-dimensional structure of the α-amylase inhibitor Tendamistat determined from NMR measurements in aqueous solution. All bonds connecting heavy atoms are shown for the residues 5-73. (Kline et al. 1988)

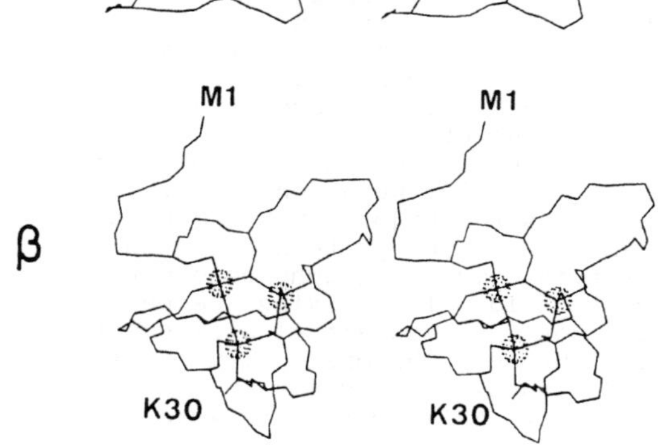

Fig. 4. Stereo view of the three-dimensional structure of the α- and β-domains of rat [Cd_7]-metallothionein-2 determined from NMR measurements in aqueous solution. The bonds between the heavy atoms of the polypeptide backbone and the metal-bound cysteine side chains are drawn, and the Cd^{2+}-ions are represented as dotted spheres. The *lettering* identifies the first and last residues in each domain. (Schultze et al. 1988)

tions i and j in the NOESY spectrum. Each NOESY cross-peak then tells us that two protons in known locations along the polypeptide chain are separated by a distance of less than approximately 5.0 Å in the three-dimensional protein structure. Since the overall length of an extended polypeptide chain with n residues is n•3.5 Å, the NOE's may thus impose stringent constraints on the polypeptide conformation. In Fig. 1C this is schematically indicated by the formation of a loop through the near approach of the two protons i' and j', which are correlated via the NOESY cross-peak k. Typically, the NOESY spectra of proteins contain many hundred cross-peaks (Fig.1A), which shows that the polypeptide chain in the three-dimensional structure contains a large number of loops of the type shown in Fig.1C. For the structural interpretation of these data, mathematical techniques, in particular distance geometry, are available to identify those three-dimensional arrangements of the linear polypeptide chain which satisfy all the experimental constraints (Braun 1987; Wüthrich 1986). Figures 2 - 4 show three structures determined using this method.

Figure 2 shows a superposition for minimal root mean square deviation (RMSD) of a group of structures for a decapeptide segment of the polypeptide hormone glucagon. Each of these structures is the result of a distance geometry calculation using the experimental NMR constraints (Braun et al. 1983). The different structures were obtained by repeating the same calculation with different, randomly generated starting conditions. The experimental constraints represent only an incomplete set of intramolecular interatomic connections, and each individual constraint describes the allowed distance range from the van der Waals contact distance of 2.0 Å to the upper limit derived from the NOE's, rather than a precise value for the distance. Therefore, the individual structures are similar but not identical. The result of a structure determination by NMR and distance geometry is commonly represented by a group of conformers, each of which represents a solution to the geometric problem of fitting the polypeptide chain to the ensemble of all experimental constraints. The spread among the different structures, usually expressed by the average of the pairwise RMSD's (Braun et al. 1983; Havel and Wüthrich 1984, 1985), gives an indication of the precision of the structure determination. Figures 3 and 4 correspond to individual conformers taken from groups of structures such as that shown in Fig.2.

The NMR method for protein structure determination is of recent origin, and its development was closely linked with that of NMR instrumentation and NMR methodology (Wüthrich 1986, 1988). The decisive advances were the introduction of NOE experiments enabling the measurement of ^1H-^1H distances in macromolecules (Gordon and Wüthrich 1978, Wagner and Wüthrich 1979; Anil-Kumar et al. 1981), the development of an efficient technique for obtaining sequence-specific assignments of the many hundred to several thousand NMR lines in a protein (Fig.1) (Dubs et al. 1979; Billeter et al. 1982; Wagner and Wüthrich 1982), and the development of distance geometry algorithms (Blumenthal 1970; Crippen 1977) capable of computing three-dimensional polypeptide structures from the NMR data (Braun and Gō 1985; Braun et al. 1981; Havel and Wüthrich 1984, 1985). Prior to the introduction of the sequential resonance assignment technique (Dubs et al. 1979; Wüthrich et al. 1982), lack of sequence-specific assignments was the major stumbling block opposing detailed analysis of the NMR spectra of proteins, and the situation was similar to that of protein crystallography before the advent of the heavy-atom derivative technique for solving the phase problem (Blundell and Johnson 1976). Both the sequential assignment technique (Dubs et al. 1979) and the three-dimensional polypeptide structure determination by NMR and distance geometry (Braun et al. 1981) were developed and applied before 2D NMR experiments were ready for the relevant measurements with proteins. However, full realization of the method was achieved only with the greatly improved resolution and efficiency afforded by 2D NMR spectroscopy (Ernst et al. 1987; Wüthrich 1986). It was therefore of great importance for the subsequent rapid progress that adaptation of existing 2D NMR experiments and the development of special new techniques for use with biomacromolecules was started immediately after the fundamental work on 2D NMR by Ernst and coworkers (Aue et al. 1976). In 1977 the first 2D NMR spectra of a protein were recorded (Nagayama et al. 1977), and by 1980 NOESY and other 2D NMR experiments could be performed in H_2O solution (Anil-Kumar et al. 1980), which provided the information needed for protein structure determinations (Wüthrich 1986). The first determination of the complete structure of a globular protein in solution was then completed in 1984 (Willamson et al. 1985).

Complementary Information from Protein Structures Determined by NMR in Solution or by Diffraction Techniques in Single Crystals

There are some aspects in protein structure determination where the complementary nature of the information obtained from X-ray diffraction and from NMR is readily apparent. Thus, since NMR can be applied to molecules for which no single crystals are available, new structures can be obtained that are not available from X-ray studies. Examples are bull seminal proteinase inhibitor IIA (Willamson et al. 1985), *E. coli lac* repressor (Zuiderweg et al. 1984) and epidermal growth factor (Montelione et al. 1987). Conversely, if the size of the protein exceeds a molecular weight of 20,000 to 30,000, or if it is not possible to prepare relatively concentrated, stable solutions, X-ray diffraction remains the only avenue to the molecular structure, provided that suitable single crystals are available.

It is also clear that NMR and X-ray diffraction yield different types of information on the internal dynamics of proteins. NMR provides direct, quantitative measurements of the frequencies of certain high activation energy motional processes, and at least semi-quantitative information on additional high frequency processes (Wagner 1983; Wagner and Wüthrich 1986). The corresponding information from X-ray structure determinations commonly consists of an outline of the conformation space covered by the combination of static disorder and high frequency structural fluctuations (Blundell and Johnson 1976). Furthermore, neutron diffraction in single crystals and NMR in solution can both be employed for studies of the exchange of labile protons, thus potentially enabling direct comparison of dynamic properties in these two states of a protein.

Since protein structure determinations by NMR or by X-ray diffraction can be performed completely independently, one now has, for the first time, a basis for meaningful comparisons of corresponding structures in single crystals and in noncrystalline states. This is highly relevant, since the solution conditions for NMR studies can often be chosen so as to coincide closely with the natural, physiological environment of the protein (Wüthrich 1986). There are presently two well-documented examples where different molecular architectures were reported for the same protein in single crystals and under nondenaturing solution conditions. One of these is the polypeptide hormone glucagon (Fig.2), which appears to have a pronounced tendency for adopting different conformations in different environments (Braun et al. 1983; Sasaki et al. 1975). Metallothionein-2 from rat liver (Fig.4) is a protein showing substantial, fundamental differences between the global molecular architectures seen in aqueous solution (Schultze et al. 1988) and in a recently published crystal structure (Furey et al. 1986). Most strikingly, 21 out of the total of 28 coordinative bonds linking the seven metal ions with the polypeptide chain (Fig.4) are different in the two reported structures.

In contrast to the forementioned examples, the α-amylase inhibitor Tendamistat (Fig.3) is a protein for which the global molecular architecture is the same in crystals and in solution (Kline et al. 1986; Pflugrath et al. 1986). Similar behavior was reported for other proteins, e.g., basic pancreatic trypsin inhibitor (BPTI) (Wagner et al. 1987) and barley serine proteinase inhibitor-2 (Clore et al. 1987), and may well turn out to be a characteristic of most globular proteins. Nonetheless, even for these proteins, detailed comparisons using refined structures in crystals and in solution, as in the case of Tendamistat (Fig.3) (Kline et al. 1988; Pflugrath et al. 1986), reveal a wealth of more subtle differences involving individual amino acid residues or groups of spatially adjacent residues. Such differences are mostly on or near the protein surface. This is not really surprising, as one would anticipate that the protein surface is most directly influenced by the protein-protein contacts in the crystal lattice. On the other hand, these differences are of special interest, since the molecular surface is generally most directly related with the functional properties of the molecule, and often includes the active site. A further extension of such comparative studies may be achieved because the solution conditions for NMR studies (e.g., pH, temperature, ionic strength, buffers) can usually be varied over a wide range, which opens new avenues for investigations of proteins under denaturing solution conditions, and for investigations of intermolecular interactions with other solute molecules (Wüthrich 1986).

Conclusion and Outlook

Considering the large pool of presently available protein crystal structures, one of the principal contributions from NMR, besides structure determinations for proteins that either cannot be crystallized or undergo major structural rearrangements during crystallization, may well turn out to consist in information toward a more meaningful interpretation of protein crystal structures under physiological solution conditions. The preceding section gives an indication that such information will come primarily from detailed studies of the protein surface. For example, one could foresee that similar patches of spatially adjacent amino acid side chains occur on the surface of different proteins, and that results from NMR studies of this surface area in a small protein could then also be applied to the other species. Such studies of the protein surface must be based on the resonance assignments and the determination of the global protein structure as outlined in Fig.1. However, because of the increased disorder and mobility near the molecular surface in protein solution structures, which is clearly seen also in Tendamistat, NMR parameters other than NOE's may play an increasingly important role in these projects (Wüthrich et al. 1982; Wüthrich 1986). Such supplementary data relating to the surface structure of a protein have in practice been obtained from amide proton titration shifts (Bundi and Wüthrich 1979) and from the combination of the two vicinal coupling constants $^3J_{\alpha\beta2}$ and $^3J_{\alpha\beta3}$ in αCH-βCH_2 fragments of amino acid side chains (Nagayama and Wüthrich 1981). Useful information may also come from many of the experiments that have been used with proteins before sequence-specific resonance assignments were independently accessible (Jardetzky and Roberts 1981; Wüthrich 1976).

Acknowledgments. The author's research has received financial support from the Schweizerischer Nationalfonds and special grants of the ETH Zürich. I thank Mr. R. Marani for the careful preparation of the typescript.

References

Anil-Kumar, Wagner G, Ernst RR, Wüthrich K (1980) Studies of J-connectivities and selective ^1H-^1H Overhauser effects in H_2O solutions of biological macromolecules by two-dimensional NMR experiments. Biochem Biophys Res Commun 96:1156-1163

Anil-Kumar, Wagner G, Ernst RR, Wüthrich K (1981) Buildup rates of the nuclear Overhauser effect measured by two-dimensional proton magnetic resonance spectroscopy: Implications for studies of protein conformation. J Am Chem Soc 103:3654-3658

Aue WP, Bartholdi E, Ernst RR (1976) Two-dimensional spectroscopy: application to NMR. J Chem Phys 64:2229-2246

Billeter M, Braun W, Wüthrich K (1982) Sequential resonance assignments in protein ^1H nuclear magnetic resonance specta. Computation of sterically allowed proton-proton distances and statistical analysis of proton-proton distances in single crystal protein conformations. J Mol Biol 155:321-346

Blumenthal LM (1970) Theory and application of distance geometry. Chelsea, New York

Blundell TL, Johnson LN (1976) Protein crystallography. Academic Press, New York

Braun W (1987) Distance geometry and related methods for protein structure determination from NMR data. Q Rev Biophys 19:115-157

Braun W, Gō N (1985) Calculation of protein conformations by proton-proton distance constraints. A new efficient algorithm. J Mol Biol 186:611-626

Braun W, Bösch C, Brown LR, Gō N, Wüthrich K (1981) Combined use of proton-proton Overhauser enhancements and a distance geometry algorithm for determination of polypeptide conformations. Application to micelle-bound glucagon. Biochim Biophys Acta 667:377-396

Braun W, Wider G, Lee KH, Wüthrich K (1983) Conformation of glucagon in a lipid-water interphase by ^1H nuclear magnetic resonance. J Mol Biol 169:921-948

Bundi A, Wüthrich K (1979) Use of amide ^1H NMR titration shifts for studies of polypeptide conformation. Biopolymers 18:299-312

Clore GM, Gronenborn AM, James MNG, Kjaer M, McPhalen CA, Poulsen FM (1987) Comparison of the solution and X-ray structures of barley serine proteinase inhibitor-2. Protein Engineering 1:313-318

Crippen GM (1977) A novel approach to calculation of conformation: distance geometry. J Comp Physics 24:96-107

Dubs A, Wagner G, Wüthrich K (1979) Individual assignments of amide proton resonances in the proton NMR spectrum of the basic pancreatic trypsin inhibitor. Biochim Biophys Acta 577:177-194

Ernst RR, Bodenhausen G, Wokaun A (1987) Principles of nuclear magnetic resonance in one and two dimensions. Clarendon, Oxford

Furey WF, Robbins AH, Clancy LL, Winge DR, Wang BC, Stout CD (1986) Crystal structure of Cd,Zn-metallothionein. Science 231:704-710

Gordon SL, Wüthrich K (1978) Transient proton-proton Overhauser effects in horse ferrocytochrome c. J Am Chem Soc 100:7094-7096

Havel TF, Wüthrich K (1984) A distance geometry program for determining the structures of small proteins and other macromolecules from nuclear magnetic resonance measurements of intramolecular ^1H-^1H proximities in solution. Bull Math Biol 46: 673-698

Havel TF, Wüthrich K (1985) An evaluation of the combined use of nuclear magnetic resonance and distance geometry for the determination of protein conformations in solution. J Mol Biol 182:281-294

Jardetzky O, Roberts GCK (1981) NMR in molecular biology. Academic Press, New York

Kline AD, Braun W, Wüthrich K (1986) Studies by ^1H nuclear magnetic resonance and distance geometry of the solution conformation of the α-amylase inhibitor Tendamistat. J Mol Biol 189:377-382

Kline AD, Braun W, Wüthrich K (1988) Determination of the complete three-dimensional structure of the α-amylase inhibitor Tendamistat in aqueous solution by nuclear magnetic resonance and distance geometry. J Mol Biol (in press)

Montelione G, Wüthrich K, Nice EC, Burgess AW, Scheraga HA (1987) Solution structure of murine epidermal growth factor: Determination of the polypeptide backbone chain-fold by nuclear magnetic resonance and distance geometry. Proc Natl Acad Sci USA 84:5226-5230

Nagayama K, Wüthrich K (1981) Structural interpretation of vincinal proton-proton coupling constants $^3J_{HH}$ in the basic pancreatic trypsin inhibitor measured by two-dimensional J-resolved NMR spectroscopy. Eur J Biochem 115:653-657

Nagayama K, Wüthrich K, Bachmann P, Ernst RR (1977) Two-dimensional J-resolved ^1H NMR spectroscopy of biological macromolecules. Biochem Biophys Res Commun 78:99-105

Pflugrath JW, Wiegand G, Huber R, Vértesy L (1986) Crystal structure determination, refinement and the molecular model of the α-amylase inhibitor HOE 467A. J Mol Biol 189:383-386

Sasaki K, Dockevill S, Ackmiak DA, Tickle IJ, Blundell TL (1975) X-ray analysis of glucagon and its relationship to receptor binding. Nature 257:751-757

Schultze P, Wörgötter E, Braun W, Wagner G, Vašák M, Kägi JHR, Wüthrich K (1988) The conformation of [Cd_7]-metallothionein-2 from rat liver in aqueous solution determined by nuclear magnetic resonance. J Mol Biol 203:251-268

Wagner G (1983) Characterization of the distribution of internal motions in the basic pancreatic trypsin inhibitor using a large number of internal NMR probes. Q Rev Biophys 16:1-57

Wagner G, Wüthrich K (1979) Truncated driven nuclear Overhauser effect (TOE). A new technique for studies of selective ^1H-^1H Overhauser effects in the presence of spin diffusion. J Magn Reson 33:675-680

Wagner G, Wüthrich K (1982) Sequential resonance assignments in protein ^1H nuclear magnetic resonance spectra. Basic pancreatic trypsin inhibitor. J Mol Biol 155: 347-366

Wagner G, Wüthrich K (1986) Observation of internal motility of proteins by nuclear magnetic resonance in solution. In: Hirs CHW, Timasheff SN (eds) Methods in enzymology 131: Enzyme structure part L. Academic Press, New York, pp 307-326

Wagner G, Braun W, Havel TF, Schaumann T, Gō N, Wüthrich K (1987) Protein structures in solution by nuclear magnetic resonance and distance geometry. The polypeptide fold of the basic pancreatic trypsin inhibitor determined using two different algorithms, DISGEO and DISMAN. J Mol Biol 196:611-639

Williamson MP, Havel TF, Wüthrich K (1985) Solution conformation of the proteinase inhibitor IIA from bull seminal plasma by ^1H nuclear magnetic resonance and distance geometry. J Mol Biol 182:295-315

Wüthrich K (1976) NMR in biological research, peptides and proteins. North Holland, Amsterdam

Wüthrich K (1986) NMR of proteins and nucleic acids. Wiley, New York

Wüthrich K (1989) The development of nuclear magnetic resonance spectroscopy as a technique for protein structure determination. Accts Chem Res (in press)

Wüthrich K, Wider G, Wagner G, Braun W (1982) Sequential resonance assignments as a basis for determination of spatial protein structures by high resolution proton nuclear magnetic resonance. J Mol Biol 155:311-319

Zuiderweg ERP, Billeter M, Boelens R, Scheek RM, Wüthrich K, Kaptein R (1984) Spatial arrangement of the three α helices in the solution conformation of *E. coli lac* repressor DNA-binding domain. FEBS Lett 174:243-247

A Calculated Conformation for the Folding Transition State of Bovine Pancreatic Trypsin Inhibitor

M. Levitt [1]

Introduction

The puzzle of how protein molecules fold from open random-coil conformations to compact precisely ordered native conformations has intrigued experimentalists and theoreticians alike ever since the native conformation of the protein myoglobin was revealed by X-ray crystallography almost 30 years ago (Kendrew et al. 1960). It is now commonly accepted that the folding process occurs along a pathway involving transitions between intermediate conformations, starting with a random coil and ending with the native conformation. Experimentally, it has proved very difficult to characterize the intermediates and transition states along the pathway, as they are all less stable than either the random coil or the native state and so never accumulate. For one small protein, BPTI, Creighton has succeeded in trapping intermediates that have different pairings of the six disulfide bonds (Creighton 1977a,b, 1978). He has analyzed the rates of interconversion of the various intermediates and so defined a pathway. The pathway turns out to be nonrandom but not to involve simple sequential formation of the three native disulfide bridges (Cys5-Cys55, Cys14-Cys38, Cys30-Cys51).

The most unexpected of his results involves the last step before the formation of the native conformation in which an intramolecular exchange of disulfide bonds is rate-limiting, occurring 1000 times more slowly than other conformational transitions. More specifically, the species with two disulfide bonds (SS_{5-14}, SS_{30-51}) undergoes an intramolecular thioldisulfide exchange reaction to give the species with two native disulfide bonds (SS_{5-55}, SS_{30-51}), which is identical in its behavior to the native conformation. For this exchange to occur, the sulfur atoms of Cys5, Cys14, and Cys55 must come into close contact, giving a transition state with a free energy observed to be 24 kcal/mol above that of the native-like species (SS_{5-55}, SS_{30-51}) (Creighton 1977b). The existence of such a transition state as the last step preceding the attainment of the native conformation is difficult to reconcile with the native conformation itself where the separation of the sulfur atoms of Cys5 and Cys14 is 23.5 Å, almost as far apart as possible in this small protein.

Method

Here energy minimization is used to generate a series of perturbated native conformations of BPTI in which Cys14 and Cys5 are forced to come progressively closer together without breaking the disulfide bond between Cys5 and Cys55. The potential energy function used before for the study of BPTI with different proline isomers (Levitt 1981b) is used.

[1]Department of Chemical Physics, The Weizmann Institute of Science, Rehovot, Israel

39. Colloquium Mosbach 1988
Protein Structure and Protein Engineering

Table 1. Energy contributions of BPTI conformations

Conformation	d_{5-14} (Å)	Energy (kcal/mol) Tot[c]	Bon	Ang	Tor	vdW	Hbon	RMS (Å)[a] All	C^{α}	R_g[b] (Å)
N	23.5	-342	3	33	-27	-228	-123	0.9	0.5	11.0
N_{14-38^-}	22.0	-6[d]	0	-2	-14	0	12	1.1	0.6	11.0
$T_{5...14}(10Å)$	10.0	0	0	-2	-5	-2	9	2.5	2.2	10.6
$T_{5...14}(5Å)$	5.1	28	1	8	3	5	12	3.7	3.4	10.5

[a]RMS is the root mean square deviation of All atoms and only the C^{α} atoms of residues 2 to 56 from the BPTI X-ray coordinates.

[b]R_g is the radius of gyration calculated as the root mean square distance of every atom (including the 61 hydrogen atoms) from the molecular centroid.

[c]Tot is the total energy, Bon is the bond stretching energy, Ang is the bond angle energy, Tor is the torsional energy, vdW is the van der Waal's energy and Hbon is the hydrogen bond energy.

[d]The energies of the last three conformations are given relative to those of the first conformation, N, the relaxed native conformation with all three disulfide bridges intact.

This function, which was derived to fit observations on small molecule spectra, molecular crystal geometries, and sublimation energies, was shown to cause a smaller deviation from the X-ray structure of BPTI than other potentials tested (Levitt 1981a; Levitt 1983).

The restraint on d_{5-14}, the distance between the sulfur atoms of Cys5 and Cys14, is implemented by adding an extra term to the molecular potential energy with the following form

$$E' = E + 100 \; [(d_{5-14} - d^{o}_{5-14})^2 \;] \text{ kcal/mol},$$

where d_{5-14} is the current distance between the sulfur atoms of Cys5 and Cys14 and d^{o}_{5-14} is the target distance to which d_{5-14} is being restrained. The total potential energy is minimized using the conjugate gradient method to simultaneously vary all 1545 Cartesian coordinates of the 515 atoms included [454 heavy atoms plus 61 peptide and amide hydrogen atoms needed for a more accurate representation of the hydrogen bond (Levitt 1981a)]. The minimization was continued until the energy value could not be further reduced (between 3000 and 5000 evaluations of the energy function).

Results and Discussion

The energy contributions, root mean square (r.m.s) deviations from the X-ray conformation, and radii of gyration of the four conformations studied here are presented in Table 1. The first conformation, known as N, is generated by unrestrained energy minimization from the X-ray coordinates; it is the relaxed native structure and acts as the reference to which the other structures are compared. The second conformation, known as N_{14-38^-}, is also generated by unrestrained energy minimization from the native structure but the SS bond between Cys14 and Cys38 is eliminated to give a native-like two-disulfide intermediate. Breaking this disulfide bond results in a small decrease in total energy and a small increase in the r.m.s deviation from the X-ray structure (see Table 1).

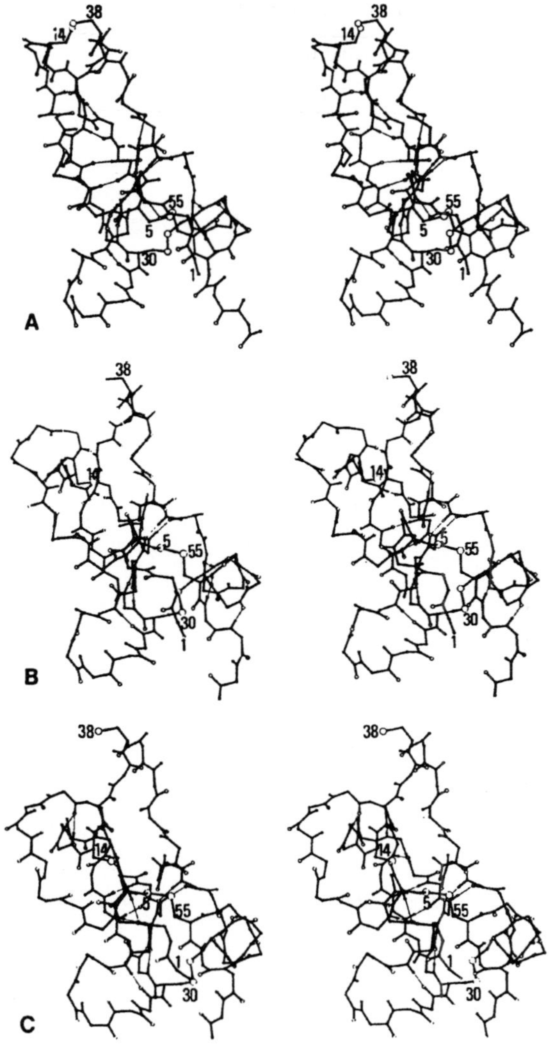

Fig. 1. Stereoscopic drawings showing the main chain, β-carbons and γ-sulfur atoms of all 58 PBTI residues viewed down the x-axis in the X-ray coordinates. Atom types are distinguished by the *size of the circle* used to represent them: the *smallest circles* are for H, N, C^{α}, and C^{β} atoms, the *next size circles* are O atoms and the *largest circles* are S^{γ} atoms. The conformations shown are as follows: (A) The minimum energy conformation known as N_{14-38}^{-}, which is closest to the X-ray conformation with one of the three native disulfide bridges, 14-38, broken. This conformation differs by 1.1 Å from the X-ray structure and has an energy that is 6 kcal/mol lower. (B) The minimum energy conformation known as $T_{5...14(10Å)}$, obtained with the restraint distance $d^{o}_{5...14}$ = 10 Å. This conformation differs by 2.5 Å from the X-ray structure and has an energy that is 6 kcal/mol higher. (C) The minimum energy conformation known as $T_{5...14(5Å)}$, obtained with the restraint distance d^{o}_{5-14} = 5 Å. This conformation differs by 3.7 Å from the X-ray structure and has an energy that is 34 kcal/mol higher

The next two conformations are generated by further minimization from N_{14-38}^{-} with target distance values d^{o}_{5-14} of 10 Å and 5 Å are known as $T_{5...14(10Å)}$ and $T_{5...14(5Å)}$, respectively. Their total energies increase as d_{5-14} is decreased; almost all of this increase being due to local strain of bond angles and torsion angles; the more global nonbonded interactions (van der Waals' plus hydrogen bonds) increase much less. Part of the higher energy associated with $T_{5...14(5Å)}$ is due to the close van der Waals' contact as S_5 and S_{14} are brought together.

It is notable that decreasing the distance between S_5 and S_{14} from 23.5 Å to 10 Å causes a small energy increase of only 6 kcal/mol. With the potential energy functions used here, this energy increase is equi-

Table 2. Comparison of main chain hydrogen bonds in BPTI conformations

Atom Pair	Hydrogen bond found in conformation[a] N_{14-38}^{-}	$T_{5..14}$(10 Å)	$T_{5..14}$(5 Å)
In X-ray			
N-terminal 3_{10}-helix			
2,O...5,N	√	-	√
3,O...6,N	√	√	-
β-hairpin			
11,O...36,N	√	-	-
16,N...36,O	-	-	-
18,N...35,O	√	√	√
18,O...35,N	√	√	√
20,N...33,O	√	√	√
20,O...33,N	√	√	√
21,N...45,O	√	√	√
21,O...54,N	√	√	√
22,N...31,O	√	√	√
22,O...31,N	√	√	√
24,N...29,O	√	√	√
24,O...28,N	√	√	√
C-terminal α-helix			
47,O...51,N	√	√	√
48,O...52,N	√	√	√
49,O...53,N	√	√	√
50,O...54,N	√	√	√
51,O...55,N	√	-	-
52,O...56,N	-	-	-
Total Broken	2	5	5
Extra hydrogen bonds			
41,O...44,N	√	√	-
2,O...6,N	-	-	√
12,O...15,N	-	-	√
12,O...16,N	-	-	√
13,O...36,N	-	√	√
51,O...56,N	-	√	√
52,O...57,N	-	√	√
Total Extra	1	4	6

[a]If the particular hydrogen bond is found in the conformation it is marked with '√'; if it is not found it is marked with a '-'. The criterion for a hydrogen bond as applied automatically is: distance from O to N < 3.3 Å, angle between H-N...O < 20°C

valent to stretching a single bond length by 0.14 Å or bending one bond angle by 18°. That the separation of between two sulfur atoms can be decreased by 13.5 Å for such a modest increase in energy supports the notion of protein conformational softness discussed before (Levitt 1974; McCammon et al. 1977).

The r.m.s. deviations from the X-ray structure also increase steadily as d_{5-14} is decreased. For the conformation $T_{5...14}$(5Å), most of the r.m.s. deviation of 3.7 Å is due to the 20 residues that move most: residues 1-5, 8-17, 25-26 and 53-58 move 4.9 Å r.m.s., whereas the other 35 residues move 2.8 Å r.m.s. It is clear from the stereoscopic drawings (see Fig. 1), that Cys14 moves more than Cys5 as they come closer together; this is to be expected as Cys5 is bonded to Cys55 while the bond between Cys14 and Cys38 has been broken. Even in $T_{5...14}$(5Å), where the two sulfur atoms have between brought closer by over 18 Å, all the main structural features of native BPTI are preserved including the N-terminal α-helix (residues 3 to 6), the β-hairpin (residues 18 to 37) and the C-terminal α-helix (residue 47 to 56). In addition, a new α-helix is seen in the region of residues 11 to 16, although it is short and rather distorted. The (ϕ,ψ) conformational angles of residues 10 to 15 are all quite close to the standard α-helical values of (-60°,-40°) as follows (-102°,-65°), (-105°,-47°), (-80°,-110°), (-71°,-23°), (-101°,4°), and (-128°,-30°).

A more detailed analysis of the conformation is afforded by the list of main chain hydrogen bonds presented in Table 2. The restrained conformations preserve 15 of the 20 hydrogen bonds found in the native conformation of BPTI (Deisenhofer and Steigemann 1975; Wlodawer et al. 1987). Four of the five broken hydrogen bonds are replaced by alternatives: 3,O...6,N becomes 2,O...6,N; 11,O...36,N becomes 13,O...36,N; and 51,O...55,N and 52,O...56,N become 51,O...56,N and 52,O...57,N, respectively. The latter pair of changes shows that the end of the C-terminal α-helix unwinds to form a π-helix (Dickerson and Geis 1969). The two additional hydrogen bonds that do not replace broken hydrogen bonds belong to the new α-helix formed in the region of residues 11 to 16 (12,O...15,N and 12,O...16,N). The preservation of ten of the 12 hydrogen bonds in the β-hairpin (Table 2, Fig.1), which bends and twists to bring Cys5 and Cys14 closer together, is further evidence of protein flexibility. The hydrogen bonds that are absent in one or other of the restrained conformations are also least stable in molecular dynamics simulations of room-temperature thermal motion about the X-ray structure (Levitt 1981c; Levitt 1983), explaining why these hydrogen bonds are easily affected by the perturbing restraints applied here.

Conclusions

The above analysis of the conformations generated by restrained energy minimization shows that the sulfur atoms of residues Cys4, Cys14, and Cys55 can be brought into close proximity in a conformation that preserves most of the favorable interactions observed in the native BPTI X-ray structure. The increase in potential energy required to decrease the distance between S_5 and S_{14} (d_{5-14}) from 23.5 Å to 5 Å is 34 kcal/mol. This energy is of similar magnitude to the free energy of 24 kcal/mol found experimentally to be associated with the transition state in the thiol-disulfide exchange reaction in which the 5-55 SS bond breaks at the same time that the 5-14 SS bond is formed. When d_{5-14} is decreased to 10 Å, the energy increase is only 6 kcal/mol, although some parts of the main chain move over 5 Å; this demonstrates the conformational flexibility of the native state.

The method of restrained energy minimization, which has also been used to compute the native conformation from the amino acid sequence (Kuntz et al. 1976; Levitt 1983), is applicable to the study of a wide range of conformational perturbations to the native conformation of a protein molecule.

Acknowledgment. This work was partially supported by a National Science Foundation award PCM-7808029. I am grateful to Leslie Orgel and Tom Creighton for critical discussion.

References

Creighton TE (1978) Prog Biophys Mol Biol 33:231-298
Creighton TE (1977a) J Mol Biol 133:295-312
Creighton TE (1977b) J Mol Biol 133:313-328
Deisenhofer J, Steigemann W (1975) Acta Crystallogr B31:238-250
Dickersen RE, Geis I (1969) The structure and action of proteins. Harper & Row, New York
Kendrew JC et al (1960) Nature 185:422-427
Levitt M (1974) In: Blout ER (ed) Peptides, proteins and polypeptides. Wiley, New York, pp 99-113
Levitt M (1981a) In: Jaenicke R (ed) Protein folding. Elsevier, Amsterdam, pp 17-39
Levitt M (1981b) J Mol Biol 145:251-263
Levitt M (1981c) Nature 294:379-380
Levitt M (1983) J Mol Biol 168:595-620
McCammon JA, Gelin BR, Karplus M (1977) Nature 267:585-590
Wlodower A, Deisenhofer J, Huber R (1987) J Mol Biol 193:145-156

Conformational Entropy and Protein Stability

K. A. Dill and D. O. V. Alonso[1]

Introduction

A principal driving force for the folding of a protein to its globular state is the tendency for the hydrophobic side chains of amino acids to cluster to avoid contact with the solvent (Kauzmann 1959). However, by itself hydrophobicity is too strong a driving force to account for the stabilities of proteins. An early estimate (Tanford 1962) suggested the hydrophobic interaction contributed the order of 180 kcal/mol toward condensation in myoglobin, but experiments (Tanford 1962; Pace 1975; Privalov and Kechinashuili 1974; Privalov 1979) show that free energies of folding are only about 1/10th as large. Hence, it is clear that there is at least one additional strong driving force in the opposite direction, toward the unfolded state, but with magnitude almost equal to that of the hydrophobic interaction. Thus, the marginal observed stabilities of globular proteins appear to be due to a small difference of large driving forces. The principal candidate for the opposing force is the conformational entropy: the conformational freedom of the molecule should clearly be greater in the many unfolded configurations than in the folded state. Statistical thermodynamic theory has recently been developed to account for this balance of forces (Dill 1985; Dill et al. 1988). Our purpose here is to focus on the role of the conformational entropy.

In order to predict the magnitude of the conformational entropy change, it is necessary to determine the number of conformations in the unfolded and folded states. An open chain molecule in solution is often well represented as a random flight. If the number of rotational states accessible to each peptide bond is q, then a chain of n bonds will have

$$\omega = q^{n-1} \tag{1}$$

accessible rotational states of the chain backbone. However, the number of accessible states changes, along with the corresponding conformational entropy, as the protein increases its local density of chain segments toward the condensed state, and as the residues become "ordered", with hydrophobic amino acids largely confined to an inner core and surrounded by a hydrophilic interfacial region. Both the condensation and ordering processes involve entropy changes, described below.

Elasticity (Short-Ranged Interactions)

According to classical polymer theory (Flory 1953), when the conformations of polymers can be suitably represented as random flights, then the number of conformations of the chain which have end-to-end lengths

---

[1]Department of Pharmaceutical Chemistry, University of California, San Francisco, CA 94143, USA

39. Colloquium Mosbach 1988
Protein Structure and Protein Engineering

between r and r + dr is

$$W(r)dr = q^{n-1}\left[\frac{3}{2\pi r_o^2}\right]^{3/2} e^{-\frac{3}{2}\left(\frac{r}{r_o}\right)^2} dr \,, \qquad (2)$$

where $r_o = bn^{1/2}$ is the r.m.s. end-to-end length of the chain, b is the bond length. The end-to-end length of the polymer is related to the radius of gyration by a constant. It is more convenient to represent this distribution in terms of the density ρ of the n chain segments distributed through the local spherical volume of radius r occupied by the protein,

$$\rho = \frac{n}{\frac{4}{3}\pi r^3} \,. \qquad (3)$$

Combinations of Eqs. (2) and (3) leads to

$$W_{elastic}(\rho) = q^{n-1}\left(\frac{1.38\rho_o}{n}\right) e^{-\frac{3}{2}\left(\frac{\rho_o}{\rho}\right)^{2/3}} \,, \qquad (4)$$

where $\rho_o = cn^{-1/2}$, where c is a constant of order 1 (Dill 1985). Using the Boltzmann-Planck equation

$$S = k \ln W \qquad (5)$$

to relate the number of accessible microscopic states to the macroscopic entropy leads to

$$\frac{\Delta S_{elastic}}{k} = -\frac{3}{2}\left[\left(\frac{\rho_o}{\rho_2}\right)^{2/3} - \left(\frac{\rho_o}{\rho_1}\right)^{2/3}\right] \,. \qquad (6)$$

This "elastic" contribution to the entropy of folding is small; the use of the gaussian approximation for purposes of illustration here probably differs little from the better Flory-Fisk approximation used in the treatments presented elsewhere (Dill 1985; Dill et al. 1988).

Excluded Volume (Long-Ranged Interactions)

The most important contribution to the entropy of folding is due to the effect of excluded volume of one chain segment by another. The distribution function, W above, accounts only for short-ranged interactions of connected neighboring residues along the chain and neglects the consequences of long-ranged interactions of residues which are spatial neighbors but are far apart along the chain sequence. If the chain density is sufficiently low, then the number of self-intersections is small and the random-flight theory approximates well the number of conformations available to real chains. But if the density of chain segments is high, then most randomly generated conformations would violate physical volume-filling constraints at points where the chain would intersect and pass through itself. In short, the "elasticity" considerations above take into account only the short-ranged interactions of nearest-neighboring residues along the chain, but real chains which are compact also have important long-ranged intrachain interactions of segments far apart in the sequence.

If the short-ranged and long-ranged interactions are independent (i.e., have additive free energies), then the effects of excluded volume can be described by a factor, ω_{steric}, which multiplies $W_{elastic}$ to give the number of accessible conformations, W, of the chain. The required excluded-volume factor is defined as the ratio of the number of conformations, Q_{excl}, which are accessible when excluded volume is taken into account divided by the number of conformations, Q_{free}, accessible when excluded volume is ignored, i.e., $\omega_{steric} = \frac{Q_{excl}}{Q_{free}}$. This ratio is calculated readily by considering the spherical volume occupied by the protein to be parceled into $m = \frac{4}{3}\pi r^3$ sites, each equal to the size of one of the n chain segments. The excluded volume contribution to the distribution function can be calculated through use of an insertion process whereby each segment is considered to be inserted, one-at-a-time into this spherical lattice of sites. The first chain segment could be inserted into any of the m sites. The second chain segment can only be inserted into any of the remaining m-1 sites. (In this mean-field argument, chain connectivity is already taken into account in the elasticity term above.) The third segment can only be inserted into any of the remaining m-2 sites. And so on, until the final insertion of chain segment n occurs into any of the remaining m-n+1 sites. Thus the total number of configurations resulting from this insertion process is

$$Q_{excl} = m(m-1)(m-2)\ldots(m-n+1) = \frac{m!}{(m-n)!} \ . \tag{7}$$

This quantity must be divided by the number of configurations accessible to the chain if excluded volume were ignored. In that case, the first segment could occupy any of the m sites. If excluded volume is neglected, then the second and all subsequent segments will also have access to any of the m sites. Hence,

$$Q_{free} = m^n \tag{8}$$

Thus, the ratio required to correct for the excluded volume contribution to the number of accessible conformation is

$$\omega_{steric} = \frac{Q_{excl}}{Q_{free}} = \frac{m!}{(m-n)!m^n} = \frac{(n/\rho)!}{(n/\rho-n)!(n/\rho)^n} \ , \tag{9}$$

where the latter equally follows from the definition $\rho = n/m$. If we use Stirling's approximation, $x! = \left(\frac{x}{e}\right)^x$, and the Boltzmann-Planck Eq. (5), then this contribution to the conformational entropy is

$$\frac{\Delta S_{excl}(\rho)}{nk} = -\left[\left(\frac{1-\rho}{\rho}\right)\ln\,(1-\rho) + 1\right] . \tag{10}$$

Note that if the chain is at low segment density, $\rho << 1$, then ω_{steric} = 1, and $\Delta S_{excl} = 0$ and effects of excluded volume are negligible. On the other hand, if the chain is at maximum density, then $\rho = 1$ and $\omega_{steric} = e^{-n}$ and $\Delta S_{excl} = -nk$. The entropy change for the density increase between these two states is large. For n = 100, $\Delta F_{excl} = -nkT \cong -60$ kcal/mol favoring the open conformations. (Note that this is not equal to the conformational entropy of folding, but only a fraction of it, because the unfolded state has a higher density than ρ_o (for details see Dill 1985; Dill et al. 1988). In contrast, the elasticity contribution is essentially independent of n, and is much smaller than that of excluded volume. This excluded volume contribution to the entropy may also be considered an osmotic pressure, since the change in segment density corresponds to a change in solvent concentration within the spherical volume occupied by the protein.

"Ordering" of the Interior/Exterior Residues

A fully condensed molecule ($\rho = 1$) can undergo further changes in entropy by changing the organization of its internal structure. The protein is in a state of disorder if hydrophobic and polar residues are randomly dispersed throughout the protein, whereas it is more ordered if hydrophobic residues are clustered into an interior region and hydrophilic residues are clustered at the solvent interface. This ordering entropy is much like an entropy of solution (Dill 1985; Dill et al. 1988). In a simple bulk mixing process, for spherical solute molecules at mole fraction x in a solvent of spherical molecules of the same size, the entropy of solution (mixing) is

$$\frac{\Delta S_{mix}}{nk} = -[x \ln x + (1-x) \ln (1-x)] \ . \tag{11}$$

Likewise, if the mole fraction of hydrophobic residues in the core of the protein is x and the mole fraction of hydrophobic residues within the surface shell is θ, then entropy of ordering is

$$\frac{\Delta S_{order}}{nk} = -f_i[x \ln x + (1-x) \ln (1-x)] \\ -f_e[\theta \ln \theta + (1-\theta) \ln (1-\theta)] \quad , \tag{12}$$

where f_i is the fraction of the m = n sites which are in the interior and f_e is the fraction of the n sites which are at the surface. This entropy of ordering can also significantly oppose folding.

Vibration

There is little reason to believe that vibrations contribute significantly to the entropy of folding. Although the folded state probably has a large vibrational entropy, there is no evidence that the vibrational entropy of the unfolded state is very different (Karplus et al. 1987).

Stiffness

It is clear from Eq.(2) that the number of accessible states increases with q, the rotational freedom around each peptide bond. Increased numbers of accessible minima in the $\Phi - \Psi$ map for the peptide bonds leads to increased q and an increased number of conformations accessible to the chain. The result of increased conformational freedom of the unfolded states is a decreased relative stability of the folded state. Hence, for example, glycines (large q) should destabilize and prolines (small q) should stabilize native states. A small experimental data base supports this prediction (Matthews et al. 1987).

Changes in these short-ranged interactions are responsible for the formation of secondary structures in proteins. Helix formation from random coils is described in this context by Zimm-Bragg theory (Zimm and Bragg 1959; Poland and Scheraga 1970; Flory 1969) and beta sheet formation has been described by Mattice and Scheraga (1984). When the formation of secondary structure accompanies condensation, then the factor q should be treated by these more refined Ising matrix methods, and q^{n-1} will be replaced by

$$q^{n-1} = [1\ 0\ 0 \cdots 0] \left[\prod_{j=1}^{n-1} U_j \right] \begin{bmatrix} 1 \\ 1 \\ 1 \\ \cdot \\ \cdot \\ \cdot \\ 1 \end{bmatrix} \tag{13}$$

where U_j is the statistical weight matrix for bond j (Flory 1969).

Some authors have used the expression

$$\frac{\Delta S}{(n-1)k} = \ln q \tag{14}$$

to compute the entropy of collapse, taking q as an adjustable parameter, but this is inappropriate in general. Collapse is driven by long-ranged interactions, rather than by short-ranged interactions envisioned in the use of this expression. For entropies of types indicated above involved in collapse, the quantity q enters in expressions for both folded and unfolded states; only entropy differences are normally of interest and they are independent of q.

Cross-Links

It was pointed out by Kauzmann (1959) that disulfide bonds and other cross-links should stabilize the globular states of proteins by reducing the conformational freedom of the unfolded states (see Fig.1). The reduction in conformational freedom depends on the length of the chain, N, between the two segments which are in contact; the theory for this dependence is originally due to Jacobson and Stockmayer (1950). From Eq.(2), it follows that the fraction of conformations in which there are two specific chain segments within a distance r and r + dr is

$$p(r) = \frac{4\pi r^2 W(r)\,dr}{q^{n-1}} = \left(\frac{3}{2\pi Nb^2}\right)^{3/2} e^{-\frac{3r^2}{2Nb^2}}\, 4\pi r^2 dr \quad . \tag{15}$$

Hence, the fraction of conformations for which two segments are within a small radius $a \ll [Nb^2]^{1/2}$ from each other is

$$\int_0^a p(r)\,dr = \left(\frac{3}{2\pi Nb^2}\right)^{3/2} \left(\frac{4\pi a^3}{3}\right) = \left(\frac{6}{\pi}\right)^{1/2} \left(\frac{a}{b}\right)^3 N^{-3/2} \tag{16}$$

Thus, the reduction of conformational freedom due to the introduction of the cross-link can be expressed as a contribution to the entropy of folding.

$$\Delta S_{x\text{-link}} = -\frac{3}{2} k \ln N + k \ln \left[\left(\frac{6}{\pi}\right)^{1/2} \left(\frac{a}{b}\right)^3\right] \quad . \tag{17}$$

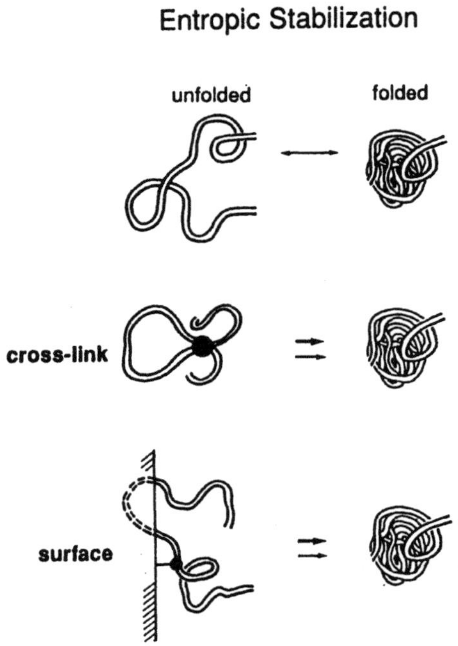

Fig. 1. Cross-links or the presence of surfaces reduce the conformational entropy of the unfolded state. Thus, they stabilize the folded state

It is clear from the sign of this expression that cross-linking decreases the entropy of the unfolded state, and to a greater extent with increasing loop size, N. Hence, the change in folding free energy due to introduction of a cross-link is negative: $\Delta\Delta F_{fold} = -(-T\Delta S_{x\text{-}link}) = T\Delta S_{x\text{-}link}$, provided the folded state is unaffected by the cross-link. Cross-linking should stabilize the folded state. There is some experimental evidence that cross-links do stabilize proteins (Perry and Wetzel 1984; Wells and Powers 1986; Wetzel 1987; Pace et al. 1988), provided the comparison is between reduced and oxidized forms of a given protein. For other experiments in which a cross-linking residue is exchanged with a non-cross-linking amino acid, there will also be other contributions to the free energy, due to different packing, etc.; in these cases, the cross-linking entropy is only one of several contributions. The effects of the addition of the cross-link cannot be so readily predicted yet in those cases.

The Jacobson-Stockmayer (JS) treatment is a random-flight theory; it does not take excluded volume into account. When effects of excluded volume are taken into account through computer simulation (Barker and Ninham 1988), it is found that the probability of loop formation depends on $N^{-23/12}$, rather than on $N^{-3/2}$ as predicted by JS (for chains in 3D). For chains in two-dimensions, where tests of the JS theory are readily performed, computer simulations and path integral methods show that stabilization also depends on the distance of the loop from the chain ends, and on interactions with other loops if more than one cross-link is present (Chan and Dill 1988). The effects of excluded volume are often more than a small correction to the cross-linking entropy given by Eq.(17); for example, some pairs of cross-links are completely prohibited by steric effects; others can be enhanced by as much as 20-fold. Hence, the Jacobson-Stockmayer theory should be considered only a qualitative predictor of effects of cross-linking.

Proteins Adjacent to Surfaces

The process of attachment of a protein to an inert, impenetrable surface should cause it to be stabilized. The presence of the surface prevents many of the unfolded conformations which could have occurred in the absence of the surface (see Fig.1). Using the reflectance principle (Feller 1957; DiMarzio 1965), we calculate the conformational stabilization due to the surface; the number of non-penetrating conformations on one side of a surface equals the total number of conformations minus those relfected through the interface. If z is the coordinate normal to the plane of the interface, and if we define

$$f(z) = \left(\frac{\beta}{\pi}\right)^{1/2} e^{-\beta z^2} ,$$

where $\beta = 3/2Nb^2$, then according to the reflectance principle, the distribution of non-penetrating conformations is

$$W(x,y,z) = \left(\frac{\beta}{\pi}\right) e^{-\beta(x^2+y^2)} \left[f(z-a) - f(z+a)\right] , \tag{18}$$

where a is the molecular "thickness" of the surface attachment region. Inasmuch as the quantity in brackets can alternatively be expressed approximately as $-2a \left.\frac{df}{dz}\right|_{z=0}$, then

$$W(x,y,z) = \left(\frac{\beta}{\pi}\right)^{5/2} e^{-\beta(x^2+y^2)} \left[4\pi aze^{-\beta z^2}\right] .$$

If the chain is attached at one end to the surface and all other segments are otherwise free (except insofar as they cannot penetrate the surface), then the fraction of the chain conformations, p(n), which remains accessible after introduction of the surface is

$$p(n) = \int_{-\infty}^{\infty} dx \int_{-\infty}^{\infty} dy \int_{0}^{\infty} dz\; w(x,y,z) = \left(\frac{6}{\pi}\right)^{1/2} \left(\frac{a}{b}\right) n^{-1/2} .$$

It follows that the entropy change in the unfolded state due to the presence of the surface is

$$\Delta S_{surf} = -\frac{1}{2} k \ln n + k \ln \left[\left(\frac{6}{\pi}\right)^{1/2} \left(\frac{a}{b}\right)\right] .$$

If the folded conformations are unaffected by the presence of the surface, then the folded state is stabilized by an amount $\Delta\Delta F_{fold} = T\Delta S_{surf}$. Similar calculations show also that further increasing the interfacial constraint by localizing and increasing the number of points of attachment should cause even greater surface-induced stabilization of the folded state.

Summary

The principal driving forces for the folding of globular proteins are entropic: the hydrophobic force favors the globular state and the conformational entropy favors the unfolded state. We provide here simple quantitative estimates for the various contributions to the conformational entropies of folding. A principal contribution is due to excluded volume, i.e., due to the osmotic pressure of solvent to enter the condensed globule. Another important contribution is due to the ordering of the residues; this disfavors a hydrophobic core and polar exterior and favors random dispersal of the residues. Both forces disfavor the folded state. The elastic entropy, which is of importance for gel swelling processes, is of lesser importance for protein folding. Important factors which can affect the conformational entropy are chain stiffness and reduction of the conformational entropy of the unfolded states by cross-linking or by the presence of inert impenetrable surfaces.

References

Barber NM, Ninham DW (1988) Random and restricted walks. Gordon and Breach, New York
Chan HS, Dill KA (1989) J Chem Phys (in press)
Dill KA (1985) Biochemistry 24:1501
Dill KA, Alonso DOV, Hutchinson K (submitted)
DiMarzio EA (1965) J Chem Phys 42:2101-2106
Feller W (1957) An introduction to probability theory and its application. Wiley, New York
Flory PJ (1953) Principles of Polymer Chemistry. Cornell University Press
Flory PJ (1969) Statistical mechanics of chain molecules. Wiley, New York
Jacobson H, Stockmayer WH (1950) J Chem Phys 18:1600
Karplus M, Ichiye T, Pettit BM (1987) Biophys J 52:1083
Kauzmann W (1959) Adv Protein Chem 14:1
Matthews BW, Nicholson H, Becktel WJ (1987) Proc Natl Acad Sci USA 84:6663-6667
Mattice WL, Scheraga HA (1984) Biopolymers 23:1701-1724
Pace N (1975) CRC Crit Rev Biochem 3:1
Pace CN, Grimsley GR, Thomson JA, Barnett BJ J Biol Chem (submitted)
Perry LJ, Wetzel R (1984) Science 226:555
Poland D, Scheraga H (1970) Theory of the helix-coil transitions in biopolymers. Academic Press, New York
Privalov PL (1979) Adv Protein Chem 33:167
Privalov PL, Kechinashville NN (1974) J Mol Biol 86:665
Tanford C (1962) J Am Chem Soc 84:4240
Wetzel R (1987) Protein Engineering 1:79
Wells JA, Powers DB (1986) J Biol Chem 261:6564
Zimm BH, Bragg JK (1959) J Chem Phys 31:526

The Improvement of Catalytic Effectiveness of an Enzyme: Pseudorevertant Triosephosphate Isomerases Obtained by Random Mutagenesis of Catalytically Sluggish Mutants

S. C. Blacklow, J. D. Hermes and J. R. Knowles[1]

Although the methods of site-directed mutagenesis now allow us to alter the amino acid residues of any cloned enzyme, and while these approaches are providing important information (Knowles 1987; Shaw 1987) that relates catalytic function more closely to protein structure, it is nevertheless true that the quality of this information is being limited by the nature of the amino acid substitutions that are being made. Thus most mutagenesis experiments are designed to test the catalytic importance of particular residues that for one reason or another are believed to participate in the catalytic act. In this sense, the method is being used to confirm existing views about structure and function, rather than to provide new and possibly surprising features of this inter-relationship. In contrast, we are making an effort to conduct an *unconstrained* search of the sequence space of an enzyme, in the hope of developing a less prejudiced view of what is structurally required, and even what is structurally ideal, for the assembly of an enzyme active site of high catalytic effectiveness.

Our starting point is triosephosphate isomerase, a glycolytic enzyme that catalyzes the chemically simple interconversion of the two triosephosphates: dihydroxyacetone phosphate and (R)-glyceraldehyde 3-phosphate. As illustrated in Fig.1, this reaction involves the intermediate formation of a cis-enediol (Rieder and Rose 1959; Knowles and Albery 1977); the enzyme merely mediates the necessary protonations and deprotonations that are required to effect the two enolization processes.

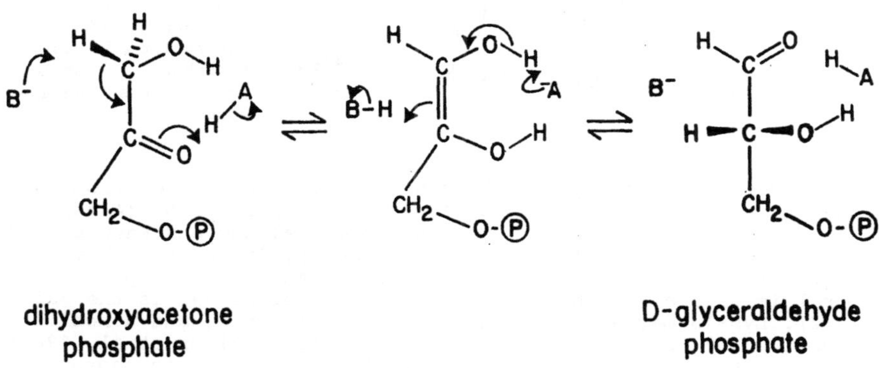

Fig. 1. The mechanistic path followed by triosephosphate isomerase. *B*- is an enzymic base and *HA* is an enzymic acid

[1]Department of Chemistry, Harvard University, 12 Oxford Street, Cambridge, Massachusetts 02138, USA

39. Colloquium Mosbach 1988
Protein Structure and Protein Engineering

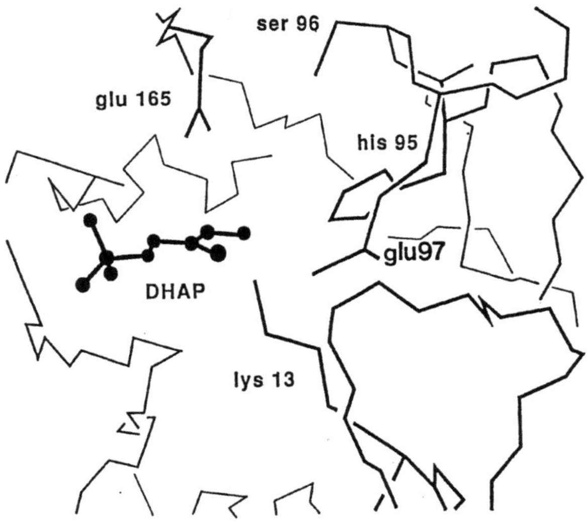

Fig. 2. The active site of triosephosphate isomerase. The substrate dihydroxyacetone phosphate has been positioned by model-building using the difference density from the low-resolution enzyme: substrate complex, the known phosphate binding site, and the known stereochemical features of the catalyzed reaction

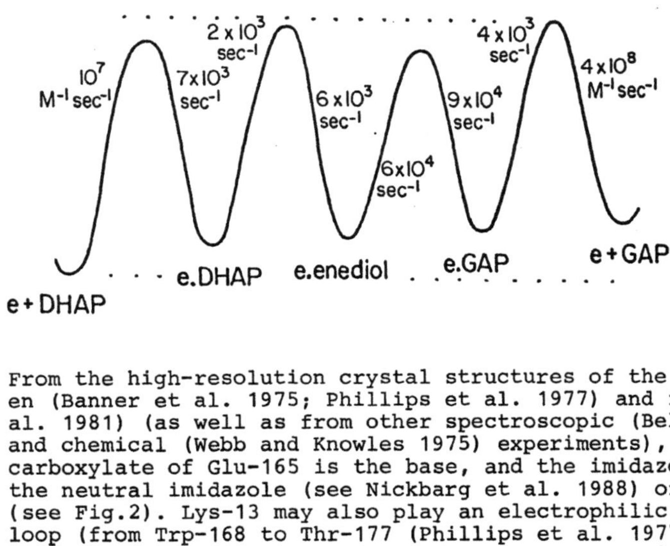

Fig. 3. The energetics of the reaction catalyzed by triosephosphate isomerase. A standard state of 40 μM (the approximate concentration of triosephosphates in muscle in vivo) has been assumed

From the high-resolution crystal structures of the isomerases from chicken (Banner et al. 1975; Phillips et al. 1977) and from yeast (Alber et al. 1981) (as well as from other spectroscopic (Belasco and Knowles 1980) and chemical (Webb and Knowles 1975) experiments), it appears that the carboxylate of Glu-165 is the base, and the imidazolium (or, perhaps, the neutral imidazole (see Nickbarg et al. 1988) of His-95 is the acid (see Fig.2). Lys-13 may also play an electrophilic role, and a mobile loop (from Trp-168 to Thr-177 (Phillips et al. 1977; Alber et al. 1981) may be important in the binding of the substrates' dianionic phospho group (Belasco et al. 1978). From a variety of kinetic experiments, the rate constants for each step of the catalyzed reaction are known (Albery and Knowles 1976a), and can be illustrated in the form of a free energy profile (see Fig.3). It should be noted that the enzyme-catalyzed reaction is very rapid: the highest transition state is that which involves the binding of the thermodynamically less stable substrate (glyceraldehyde phosphate) to the enzyme. The rate constant for this process is 4×10^8/M/s, and we have recently shown by viscosity variation experiments (Blacklow et al. 1988) that the reaction catalyzed by the wild-type enzyme is, indeed, limited only by the rate at which the substrate encounters the enzyme (Albery and Knowles 1976b).

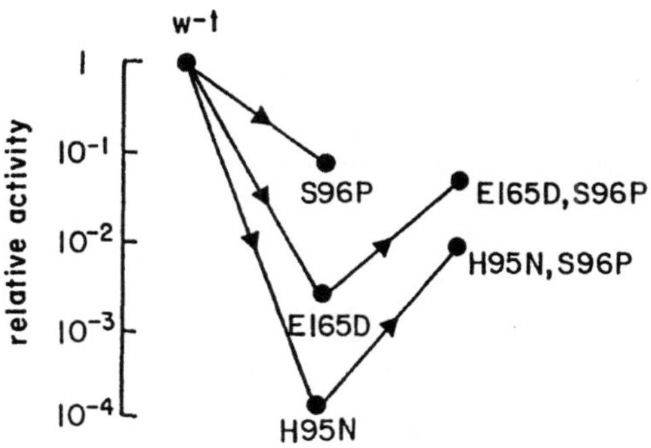

Fig. 4. Relative values of k_{cat}/K_m of various mutant triosephosphate isomerases, and of pseudorevertants of improved catalytic potency derived therefrom

In an early attempt to improve our understanding of this enzymic reaction, we changed the active site base to aspartate, to produce the E165D mutant (Straus et al. 1985). This change, which should result in a movement of the critical carboxylate by less than 1 Å (at least in principle, though we still have no crystal structure of the mutant), produces an isomerase that has a value of k_{cat}/K_m nearly 500 times lower than the wild-type enzyme (Straus et al. 1985). When the kinetic parameters for the mutant enzyme were deconvoluted (Raines et al. 1986), it became clear that the two enolization steps had each been slowed by about 1000-fold, largely because of a destabilization of the transition states for these steps. The proper interpretation of these large kinetic effects must, of course, await detailed structural information on the E165D enzyme. But since we now have a catalytically inefficient enzyme that is certainly not limited in rate by substrate diffusion, we may use it as a starting point in the search for new pseudorevertant enzymes that show improved catalytic behavior. Accordingly, we have taken the gene encoding the E165D mutant enzyme, and have subjected it to random chemical mutagenesis, following the protocols suggested by Myers et al. (1985). The fortunate existence of a selection procedure (Hermes et al. 1988) has then allowed us to find a pseudorevertant enzyme of much improved kinetic properties (see Fig.4). This enzyme carries a second mutation in which Ser-96 has become a proline, and is therefore designated E165D,S96P. While we shall here refrain from any detailed structural interpretation of this amino acid replacement, we can ask how the energetics have changed to produce a functionally better catalyst. The kinetic results show that this second-site change to produce E165D,S96P causes the enzyme to bind all substrate species more tightly. Thus the K_m for dihydroxyacetone phosphate is reduced by 30-fold (implying stronger binding of substrate by the enzyme), yet k_{cat} for this substrate is essentially unchanged (showing that the corresponding transition state is also more strongly bound by an equal factor).

This success in the search for second-site mutations that compensate, at least partially, for the original catalytic lesion (in E165D), prompted us to ask if such changes are rare events, or whether there exists a plurality of ways to improve the catalytic effectiveness of a sluggish enzyme. Accordingly, we repeated the chemical mutagenesis and search for pseudorevertants from the slow E165D isomerase. Yet the only enzyme of improved catalytic potency that emerged from this second attempt was again E165D, S96P (J.D. Hermes and W. Downs, unpublished experiments).

Concomitantly with these experiments, we investigated a primary mutation of His-95 (S.C. Blacklow, unpublished experiment), which is a strong candidate for the electrophilic component of catalysis (see Fig.2). The mutant H95N was found to be less active than the wild-type enzyme by more than 1500-fold (Fig.4). While a detailed study of the catalytic and energetic characteristics of the analogous mutant H95Q has shown a subtle but significant change in reaction mechanism (Nickbarg et al. 1988), the H95N mutant isomerase provides a second attractive starting point for random mutagenesis experiments. We therefore subjected the gene encoding the H95N enzyme to chemical mutagenesis according to Myers et al. (1985) and searched for second-site suppressor mutants as before. We were gratified quickly to find a mutant having 60 times the activity of the H95N starting material (S.C. Blacklow, unpublished experiments; Hermes et al. 1987). This pleasure diminished, however, when the second-site suppressor was found again to be S96P! That is, the S96P change partially compensates for *each* of the primary deleterious mutations, E165D and H95N (see Fig.4). There are two obvious ways to account for these findings. First, it could be the case that S96P is some kind of global suppressor, in loose analogy with those found by Shortle and his group in their work on the thermal stability of mutants of *Staphylococcal* nuclease (Shortle and Lin 1985). It is, of course, hard to imagine how, at the molecular level, a single mutation (of S96P) could partially compensate for two such different changes in the catalytic apparatus as E165D and H95N. So we were hardly surprised to find that the single S96P change in the wild-type enzyme lowers the catalytic activity by about tenfold (Fig.4). The replacement of Ser-96 by Pro is clearly *not* always beneficial in catalytic terms. The second possibility is that the S96P second-site changes are effective catalytically for different reasons, and that there is a "hot spot" for chemical modification of the gene at this site. The existence of such hot spots in mutagenesis experiments is well known (e.g., Coulondre et al. 1978), and could severely limit the range of changes that is scanned by chemical mutagenesis. In order to avoid such arbitrary and undefined constraints on the mutations that are generated, we have started to evaluate "spiked" oligonucleotide primers to produce definably random mutations over the whole length of the triosephosphate isomerase gene (Hermes and Knowles 1988).

In order to be certain that these spiked primers would indeed lead to appropriately random mutations, more than 100 mutants were picked (without any functional selection) and the relevant region of the gene was sequenced. In this way, we showed that it is possible to produce oligonucleotide primers that contain the statistically predicted number, type, and distribution of alterations in the isomerase gene (Hermes and Knowles 1988). The region of "sequence space" that we are searching is therefore defined. In the first experiment, so as to provide a direct comparison between the chemical mutagenesis approach (Myers et al. 1985) and our spiked primer method (Hermes and Knowles 1988), we have chosen to search once more for pseudorevertants from the E165D mutant isomerase. Although these first experiments are still at an early stage, we have already found our early E165D,S96P second-site suppressor, *and* a new different second-site suppressor mutant which we have characterized as E165D,E97D (J.D. Hermes, D.L. Pompliano and S.C. Blacklow, unpublished experiments). This finding already suggests that the spiked oligonucleotide primer method is allowing a more extensive search of catalytically beneficial amino acid substitutions in the sluggish E165D enzyme.

These findings augur well for the realization of our goal to select for functional improvement and to learn the importance of particular structural determinants. The mechanistic and energetic characterization of these pseudorevertants will allow us not only to map the accessible range of second-site suppressor mutants, but also to achieve further improvements in the catalytic efficiency of the enzyme by changes at third and higher-order sites.

Acknowledgments. This work was supported by the National Institutes of Health, and Merck, Sharp & Dohme. We gratefully acknowledge the continuing collaboration with Milligen Corporation on the spiked oligonucleotide method.

References

Alber T, Banner DW, Bloomer AC, Petsko GA, Phillips DC, Rivers PS, Wilson IA (1981) On the three-dimensional structure and catalytic mechanism of triosephosphate isomerase. Philos Trans R Soc Lond B293:159-171

Albery WJ, Knowles JR (1976a) Free energy profile for the reaction catalyzed by triosephosphate isomerase. Biochemistry 15:5627-5631

Albery WJ, Knowles JR (1976b) Evolution of enzyme function and the development of catalytic efficiency. Biochemistry 15:5631-5640

Banner DW, Bloomer AC, Petsko GA, Phillips DC, Pogson CI, Wilson IA, Corran PH, Furth AJ, Milman JD, Offord RE, Priddle JD, Waley SG (1975) Structure of chicken triosephosphate isomerase determined crystallographically at 2.5 Å resolution using amino acid sequence data. Nature Lond 255:609-614

Belasco JG, Knowles JR (1980) Direct observation of substrate distortion by triosephosphate isomerase using Fourier transform infrared spectroscopy. Biochemistry 19:472-477

Belasco JG, Herlihy JM, Knowles JR (1978) Critical ionization states in the reaction catalyzed by triosephosphate isomerase. Biochemistry 17:2971-2978

Blacklow SC, Raines RT, Lim WA, Zamore PD, Knowles JR (1988) Triosephosphate isomerase catalysis is diffusion controlled. Biochemistry 27:1158-1167

Coulondre C, Miller JH, Farabaugh PJ, Gilbert W (1978) Molecular basis of base substitution hotspots in *Escherichia coli*. Nature Lond 274:775-780

Hermes JD, Knowles JR (1988) The search for catalytic improvement in an enzyme: the generation of and the selection from pools of random mutants. Proceedings Robert A Welch Foundation Conferences on Chemical Research XXXI. Design of Enyzmes and Enzyme Models, pp 207-226

Hermes JD, Blacklow SC, Gallo KA, Bauer AJ, Knowles JR (1987) The evolution of enzyme function. In: Oxender DL (ed) Protein structure, folding and design, 2. UCLA Symposia on Molecular and Cellular Biology. AR Liss, New York, pp 257-264

Hermes JD, Blacklow SC, Knowles JR (1988) The development of enzyme catalytic efficiency: an experimental approach. Cold Spring Harbor Symposia on Quantitative Biology 52:597-602

Knowles JR (1987) Tinkering with enzymes: what are we learning? Science 236:1252-1258

Knowles JR, Albery WJ (1977) Perfection in enzyme catalysis: the energetics of triosephosphate iosmerase. Acc Chem Res 10:105-111

Myers RM, Lerman LS, Maniatis T (1985) A general method for saturation mutagenesis of cloned DNA fragments. Science 229:242-247

Nickbarg EB, Davenport RC, Petsko GA, Knowles JR (1988) Triosephosphate isomerase from yeast: removal of a putatively electrophilic histidine results in a subtle change in catalytic mechanism. Biochemistry 27:5948-5960

Phillips DC, Sternberg MJE, Thorton JM, Wilson IA (1977) An analysis of the three-dimensional structure of chicken triosephosphate isomerase. Biochem Soc Trans 5:642-647

Raines RT, Sutton EL, Straus DR, Gilbert W, Knowles JR (1986) Reaction energetics of a mutant triosephosphate isomerase in which the active site glutamate has been changed to aspartate. Biochemistry 25:7141-7154

Rieder SV, Rose IA (1959) The mechanism of the triosephosphate isomerase reaction. J Biol Chem 234:1007-1010

Shaw WV (1987) Protein engineering: the design, synthesis and characterization of factitious proteins. Biochem J 246:1-17

Shortle D, Lin B (1985) Genetic analysis of staphylococcal nuclease: identification of three intragenic "global" suppressors of nuclease-minus mutations. Genetics 110:539-555

Straus D, Raines RT, Kawashima E, Knowles JR, Gilbert W (1985) Active site of triosephosphate isomerase: in vitro mutagenesis and characterization of an altered enzyme. Proc Natl Acad Sci USA 82:2272-2276

Webb MR, Knowles JR (1975) The orientation and accessibility of substrates at the active site of triosephosphate isomerase. Biochemistry 14:4692-4698

Reaction Coordinate Approach to the Binding of Ligands to Carboxypeptidase A

D. W. Christianson[1] and W. N. Lipscomb[2]

The metalloenzyme carboxypeptidase A (CPA; peptidyl-L-amino acid hydrolase, EC 3.4.17.1) is an exopeptidase of molecular weight 34,472 containing one zinc ion bound to a single polypeptide chain of 307 amino acids (Quiocho and Lipscomb 1971; Hartsuck and Lipscomb 1971; Bradshaw et al. 1969). Its biological function is the hydrolysis of C-terminal amino acids from polypeptide substrates, and it prefers those substrates bearing large hydrophobic C-terminal side chains such as phenylalanine. Important residues for binding and catalysis are Glu-270, Arg-71, Arg-127, Asn-144, Arg-145, Tyr-198, Tyr-248, the active site zinc ion and its bound solvent molecule. Important details regarding the hydrolytic mechanism remain unclear and have been the subject of various reviews (Lipscomb 1980,1982,1983; Vallee et al. 1983; Vallee and Galdes 1984). The enzyme probably engages a promoted-water mechanism toward the hydrolysis of peptide substrates, but there is evidence that catalysis may proceed through an acylenzyme intermediate (a mixed anhydride with the gamma-carboxylate of Glu-270) in the hydrolysis of certain ester substrates (Makinen et al. 1979; Suh et al. 1985; Sander and Witzel 1985).

X-ray crystallographic methods have been applied toward the resolution of these details, and structural studies have focused on the interaction of the enzyme with different ligands. Among enzyme-bound ligands it is important to note the factors distinguishing substrate, transition state, reaction coordinate, and/or product analogs. Each type of inhibitor can provide valuable information regarding the association of enzyme with substrate. However, only transition state analogs (Wolfenden 1972,1976; Lienhard 1972,1973), and even more so, reaction coordinate analogs, can provide direct structural inferences pertaining to the catalytic conformations of the enzyme. Since the transition state is thought to bind to an enzyme much more tightly than substrate (Pauling 1946), it is not surprising that chemically stable analogs of transition states often display K_i values which are orders of magnitude greater than K_m values for corresponding substrates (Wolfenden 1972,1976; Lienhard 1972,1973).

Before a discussion of CPA-ligand binding can commence, we must first define the new term reaction coordinate analog. A reaction coordinate analog is a chemically reactive substrate analog. It can undergo a reversible chemical reaction identical to the first elementary step(s) of catalysis, yet it cannot complete the chemistry of the entire catalytic cycle due to a subsequently insurmountable barrier. Reaction coordinate analogs differ from transition state analogs in that the latter type of enzyme inhibitor is chemically inert, whereas the former is chemically reactive. Moreover, since the chemical reactivity of the reaction coordinate analog parallels the reversible reactivity of substrates, it cannot be classified as an irreversible mechanism-based or suicide enzyme inhibitor. Reaction coordinate analogs are perhaps

[1]Department of Chemistry, University of Pennsylvania, Philadelphia, PA 19104-6323, USA

[2]Gibbs Chemical Laboratories, Department of Chemistry, Harvard University, Cambridge, MA 02138, USA

39. Colloquium Mosbach 1988
Protein Structure and Protein Engineering

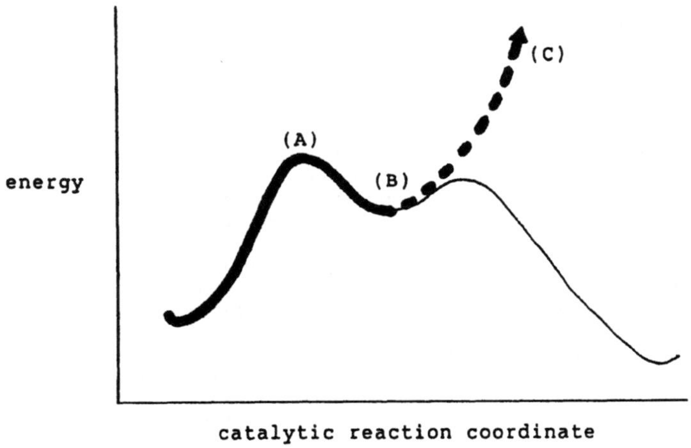

Fig. 1. An inert transition state analog mimics the structure of only one point along a given reaction coordinate, optimally that of a chemical transition state (A). A reaction coordinate analog, however, is a reactive substrate analog which undergoes a reversible chemical transformation parallel to that of an actual substrate in the first elementary step of the catalytic reaction coordinate (B). The reaction coordinate analog cannot complete the chemistry of the catalytic cycle, but must stop at some intermediate point along the reaction coordinate due to an inherent barrier to subsequent reactivity (C) [e.g., expulsion of hydride anion by a tetrahedral gem-diol(ate)]. The observation of the intermediate point (B), and the implication of chemistry leading to (B), allows the X-ray crystallographer to draw meaningful conclusions regarding mechanism. We emphasize that the reaction coordinate analog is not an irreversible mechanism-based or suicide inhibitor. Such compounds undergo noncatalytic chemistry when bound to enzyme active sites

the best candidates for X-ray study because their binding invokes the same reversible chemistry which occurs during catalysis (Fig.1) – the X-ray crystallographer can study this chemistry in great structural detail.

A review of CPA-ligand complexes logically proceeds along the entire reaction coordinate of proteolysis, and begins with an intact pseudosubstrate and a substrate analog. Then, four reaction coordinate analogs and one transition state analog may resemble a chemical intermediate (or its flanking transition states). The chemistry involved in the binding of reaction coordinate analogs probably parallels that of the first step of proteolysis. Finally, the hydrolysis products of proteolysis and phosphonamidolysis are studied as bound to the CPA active site. The study of mechanism by the structural correlation of a group of enzyme-inhibitor complexes is similar to the reaction coordinate approach toward the study of small molecule reactivity set forth by Bürgi and Dunitz (1983).

The pseudosubstrate glycyl-L-tyrosine (Fig.2) binds to CPA at -9°C as the intact peptide, but both the peptide carbonyl oxygen and the amino terminus (as the free base) coordinate to the active site zinc ion in a chelate interaction (Christianson and Lipscomb 1986a). The zinc-carbonyl interaction displayed in this complex has substantiated the proposal for a similar, productive interaction in the mechanism of peptide hydrolysis. This interpretation, however, is questionable, since glycyl-L-tyrosine binds nonproductively, having excluded the

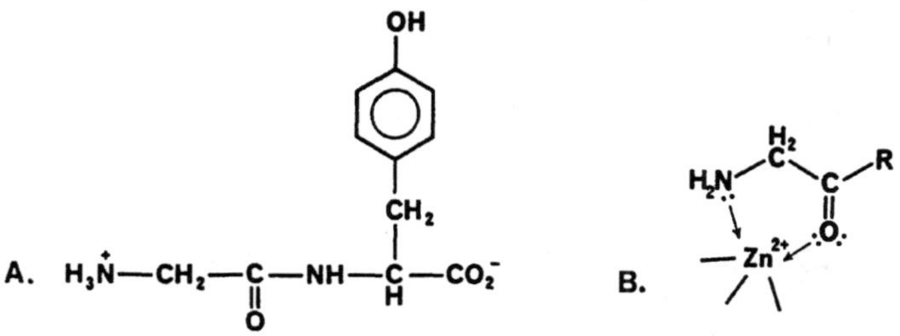

Fig. 2. The slowly hydrolyzed substrate glycyl-L-tyrosine (A) is a competitive inhibitor of rapidly-hydrolyzed CPA substrates. Although depicted here as the zwitterion, X-ray studies show that it binds nonproductively to CPA via a chelate interaction with the scissile carbonyl and the terminal amino group as the free base (B)

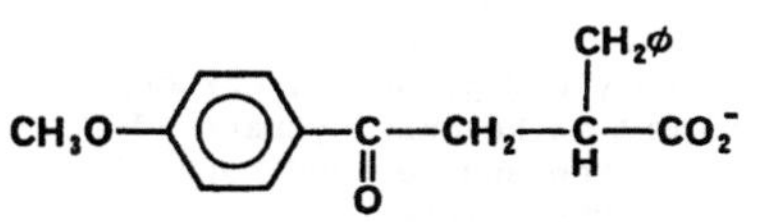

Fig. 3. The ketone (--)-3-(p-methoxybenzoyl)-2-benzylpropanoic acid inhibits CPA with a K_i of 1.1×10^{-4} M

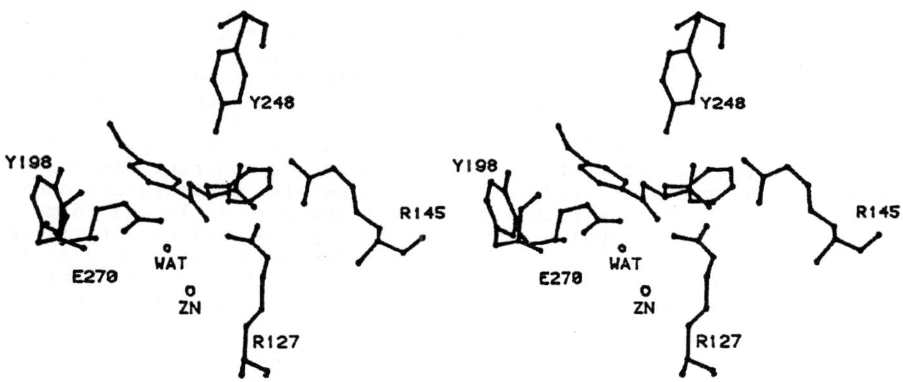

Fig. 4. Stereoview of the complex between CPA and (--)-3-(p-methoxybenzoyl)-2-benzylpropanoic acid. The inhibitor binds with its carbonyl hydrogen bonded to Arg-127. This binding mode may provide a model for the binding of peptide substrates prior to catalysis

catalytically required zinc-bound water molecule from the active site. Accordingly, glycyl-L-tyrosine is hydrolyzed nearly 5000 times more slowly than N-protected dipeptides (Quiocho and Lipscomb 1971; Izumiya and Uchio 1959). It has been confirmed in chemical studies that glycyl-L-tyrosine binds to the enzyme as the anion rather than the zwitterion (Yanari and Mitz 1956).

Perhaps a better representation of enzyme-substrate interaction can be visualized in the complex of CPA with the substrate analog 3-(p-methoxybenzoyl)-2-benzylpropanoic acid (Fig.3) (Sugimoto and Kaiser

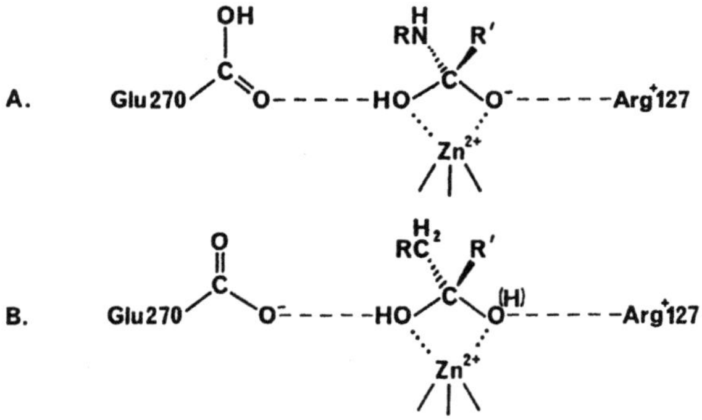

Fig. 5. The first step of the CPA-catalyzed proteolytic mechanism may be the direct attack of water, promoted by zinc and assisted by Glu-270, at the scissile peptide carbonyl (A). The resulting tetrahedral intermediate subsequently collapses, with obligatory proton donation to the leaving amino group, to yield products. Carbonyl-containing reaction coordinate analogs (Fig.6) bind to CPA as ketonic hydrate adducts (B). The chemistry of hydration may be performed by the enzyme, and both zinc and Arg-127 may help to polarize the bound inhibitors as gem-diolates

1978,1979). This ketone binds with an intact carbonyl hydrogen bonded to the guanidinium moiety of Arg-127, and it does not perturb the coordination polyhedron of the native zinc ion (Fig.4) (Christianson et al. 1985). The zinc-bound solvent molecule is 3.1 Å from the ketone carbonyl and appears to be in good orientation for attack at the π^* orbital of the carbonyl system. We believe that this enzyme-inhibitor complex provides an excellent model for the Michaelis complex of peptide substrates with CPA. The precatalytic binding of peptides to Arg-127 is consistent with various chemical data: K_m for peptide substrates is not metal-dependent, implying that a metal-carbonyl interaction in the precatalytic Michaelis complex is not apparent (Auld and Holmquist 1974); peptides bind to the apoenzyme with dissociation constants similar to their K_m values with the holoenzyme (Auld and Holmquist 1974), and X-ray studies show that the peptide carbonyl of glycyl-L-tyrosine binds to Arg-127 in the apoenzyme (Rees and Lipscomb 1983); and likewise, peptides bind to the substitution-inert Co(III)-substituted enzyme, presumably with the carbonyl bound to a nonmetal site (Van Wart and Vallee 1978).

The binding of the scissile peptide carbonyl to Arg-127 allows for subsequent attack by zinc-bound water. The nucleophilic promotion of the water molecule could be effected by zinc and assisted by Glu-270. In order to study the structural implication of this first elementary step of catalysis, the complexes of CPA with four reaction coordinate analogs have been investigated by X-ray diffraction methods. An aldehyde (Galardy and Kortylewicz 1984), an α-trifluoroketone (Gelb et al. 1985), and two ketonic substrate analogs (Grobelny et al. 1985; Oren et al. 1988) have been studied. In each case, the enzyme selects the highest-affinity form of the inhibitor, and that form is the covalent hydrate adduct (i.e., with a gem-diol function at the former carbonyl carbon) (Fig.5). The question has not yet been addressed via solution methods whether CPA chemically participates in the hydration reaction, but the expectation of this chemistry is realistic in light of the known activity of carbonic anhydrase, a related zinc enzyme, toward carbonyl hydration (Pocker and Sarkanen 1978). This reversible behavior parallels the first elementary step of proteolysis, so carbonyl-containing inhibitors (with isosteric peptide-to-ketone carbonyls) can

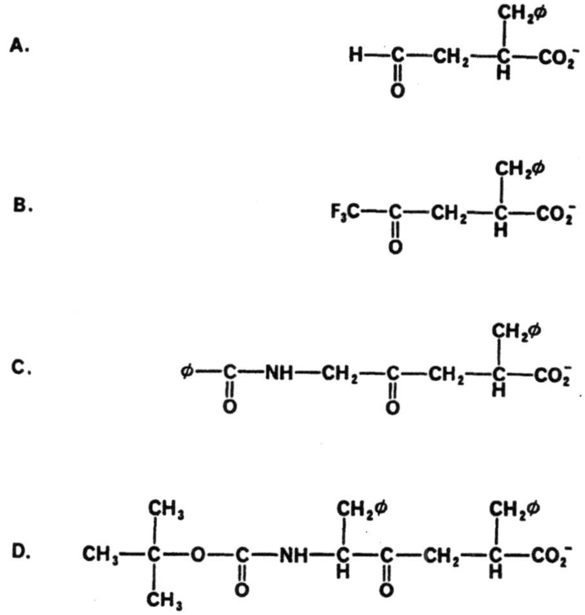

Fig. 6. Reaction coordinate analogs of the CPA-catalyzed proteolytic mechanism. (A) 2-benzyl-3-formylpropanoic acid inhibits CPA with a K_i of 4.8×10^{-7} M (enantiomeric mixture). (B) 2-benzyl-4-oxo-5,5,5-trifluoropentanoic acid inhibits CPA with a K_i of 2×10^{-7} M (enantiomeric mixture). (C) 5-benzamido-2-benzyl-4-oxopentanoic acid inhibits CPA with a K_i of 4.8×10^{-5} M (enantiomeric mixture). (D) 5-amino-(N-tert-butoxycarbonyl)-2-benzyl-4-oxo-6-phenylhexanoic acid inhibits CPA with a K_i of 6.7×10^{-7} M (diastereomeric mixture)

be classified as reaction coordinate analogs. We point out that a nucleophilic hydrolytic pathway (i.e., via a mixed anhydride intermediate with Glu-270) cannot be excluded for certain ester substrates (Makinen et al. 1979; Suh et al. 1985; Sander and Witzel 1985) on the basis of structural studies described herein. However, we might be able to address such questions if reaction coordinate analogs of those specific ester substrates were synthesized and assayed for activity with CPA.

The aldehyde 2-benzyl-2-formylpropanoic acid (Fig.6A) binds to CPA as the hydrate (Christianson and Lipscomb 1985), as does the highly electrophilic α-trifluoroketone 2-benzyl-4-oxo-5,5,5-trifluoropentanoic acid (Fig.6B) (Christianson and Lipscomb 1986b). This is not surprising in view of the enhanced electrophilicity of these compounds. It is surprising, however, to find that the non-activated ketones 5-benzamido-2-benzyl-4-oxopentanoic acid (Fig.6C) (Christianson et al. 1987) and 5-amino-(N-tert-butoxycarbonyl)-2-benzyl-4-oxo-6-phenylhexanoic acid (Figs.6D,7) (Shoham et al. 1988) bind to the enzyme as the tetrahedral hydrates. The latter pair of compounds are not very electrophilic at their ketonic carbonyls, and in solution they must exist in hydrate form to less than 0.2%. We arrive at this number because the simple ketone acetone exists in solution as 0.2% hydrate form (Lewis and Wolfenden 1977); this percentage will surely decrease for more complex ketonic centers. The enzyme may well have performed a hydration reaction on the intact carbonyls of the ketones to result in the gem-diol(ate) analogs of the proteolytic tetrahedral intermediate (Fig.5). The polarization of one hydrate oxygen by zinc probably stabilizes that oxygen as a zinc-bound anion; therefore, the observed structure may bear both a steric and electronic resemblance to the actual proteolytic tetrahedral intermediate. As previously noted, the enzyme may participate in the chemistry of hydration. Alternatively, the enzyme may have sequestered the pre-formed hydrate, as it was maintained at equilibrium in the solution in which CPA crystals were soaked, even though the con-

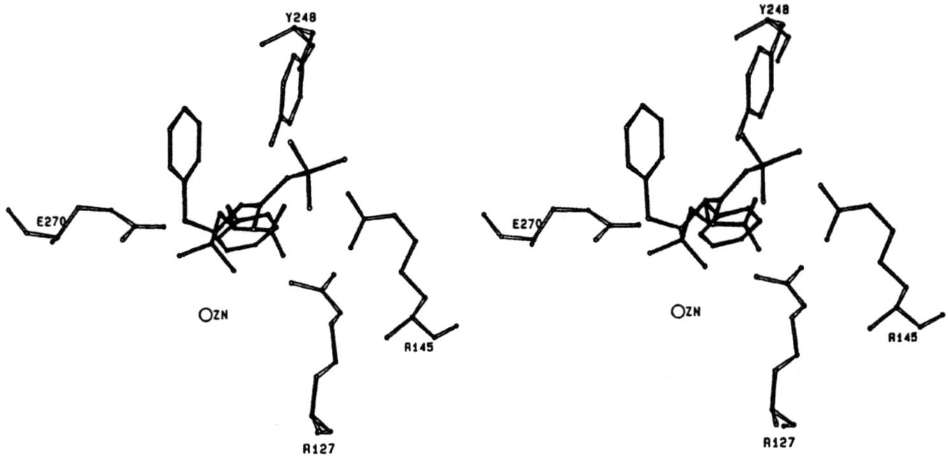

Fig. 7. Stereoview of the complex of CPA with the hydrate form of the reaction coordinate analog 5-amino-(N-tert-butoxycarbonyl)-2-benzyl-4-oxo-6-phenylhexanoic acid (Fig.6D). (Shoham et al. 1988)

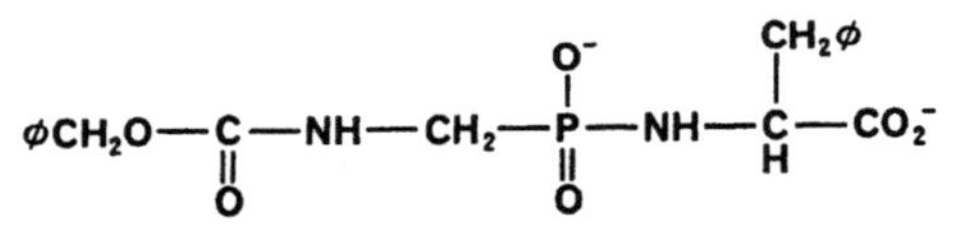

Fig. 8. The intact phosphonamidate N-[[[(benzyloxycarbonyl)amino]methyl]hydroxyphosphinyl]-L-phenylalanine inhibits CPA with a K_i of 9×10^{-8} M

centration of hydrate form was certainly less than 0.2%. We include this caveat pending the results of a chemical study of CPA-catalyzed carbonyl hydration reactions. Upon chemical confirmation, then, we classify the carbonyl-containing inhibitors of Fig.5 as reaction coordinate analogs of the CPA-catalyzed proteolytic mechanism.

In contrast with the reaction coordinate analogs, the presumably inert phosphonamidate inhibitor N-[[[(benzyloxycarbonyl)amino]methyl]hydroxyphosphinyl]-L-phenylalanine (Fig.8) (Jacobsen and Bartlett 1981), a pre-formed transition state analog of the peptide substrate Cbz-Gly-Phe, has been studied in its complex with CPA (Christianson and Lipscomb 1986c,1988). Surprisingly, this static analog is itself a "substrate" for hydrolysis when bound to the enzyme active site at pH 7.5. We make this conclusion on the basis of X-ray crystallographic observations. At pH 7.5 the hydrolyzed phosphonamidate is observed, i.e., S_2-S_1 Cbz-Gly-phosphonate plus S_1' phenylalanine; at pH 8.5, the intact phosphonamidate is found. Although CPA-phosphonamidase activity was not detected in solution with this inhibitor (Jacobsen and Bartlett 1981), we must conclude that the hydrolysis did in fact occur at the enzyme's active site over the time course of X-ray data collection (ca. 8 days). A definition for "catalysis" within these circumstances, however, is not clear. The intact binary complex represents a possible static transition state analog bound to the enzyme, as does the cleaved ternary complex with the tetrahedral phosphonate bound to zinc. However, the ternary product complex with a P_1 anion and P_1' phenylalanine may also

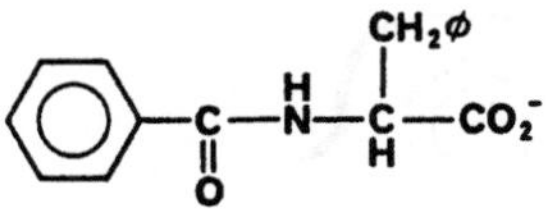

Fig. 9. The amide substrate N-benzoyl-L-phenylalanine, a slowly hydrolyzed CPA substrate. The reaction barrier to hydrolysis of this compound may be reflected in the barrier to hydration of the corresponding ketonic analog (Fig.3) which binds to CPA with an intact carbonyl

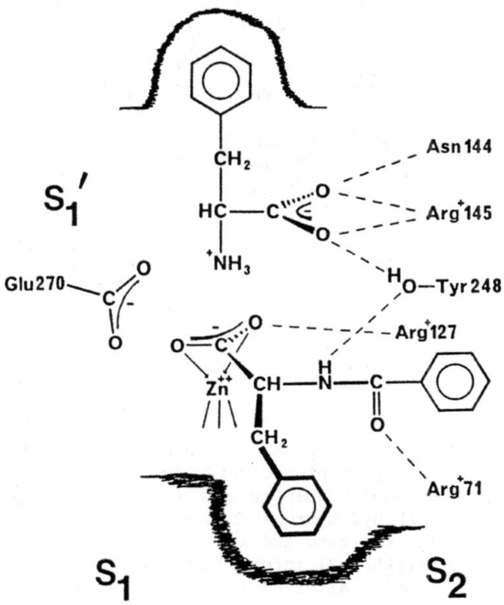

Fig. 10. CPA crystals soaked in a buffer solution containing 0.1 M benzoyl-L-phenylalanine show enzymic activity during the time frame of the X-ray diffraction experiment. Here, a schematic view of the novel enzyme-substrate-product complex is presented; note that this could also represent the enzyme-product complex for the hydrolysis of N-benzoyl-L-phenylalanyl-L-phenylalanine

mimic, at least in part, the binding of proteolytic products to the enzyme. Another ternary system, a novel enzyme-substrate-product complex of CPA with N-benzoyl-L-phenylalanine (Fig.9) and phenylalanine observed by X-ray crystallography (Christianson and Lipscomb 1987), may also represent the binding of products. In particular, the binding of S_2-S_1 N-benzoyl-L-phenylalanine and S_1' phenylalanine would be expected for the hydrolysis products of the peptide substrate N-benzoyl-L-phenylalanyl-L-phenylalanine (Fig.10). We find in each case that the zinc-bound anion (intact phosphonamidate, product phosphonate, product carboxylate) also makes a contact with Arg-127. Hence, this versatile residue is now implicated in the binding of product carboxylate, as well as substrate and tetrahedral intermediate. Residue Glu-270 is observed to make a close contact with the other oxygen of the zinc-bound anion in these examples, so we postulate that Glu-270 may be un-ionized to avoid unfavorable electrostatic interaction with the zinc-bound anion.

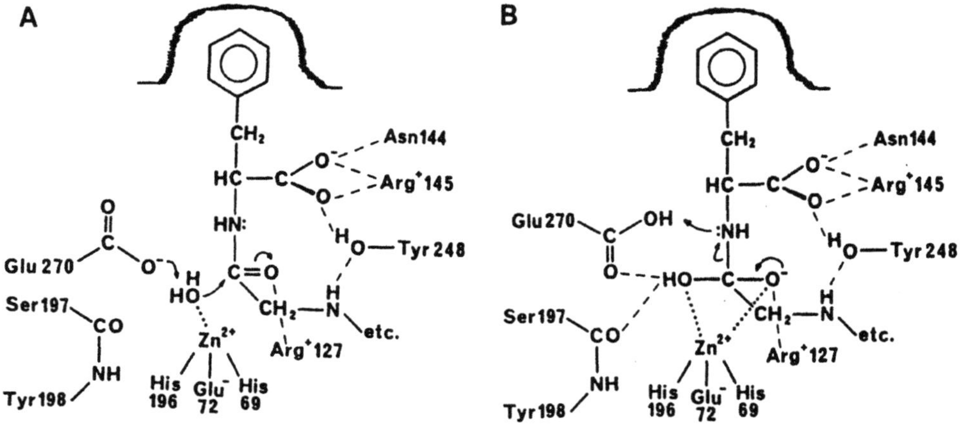

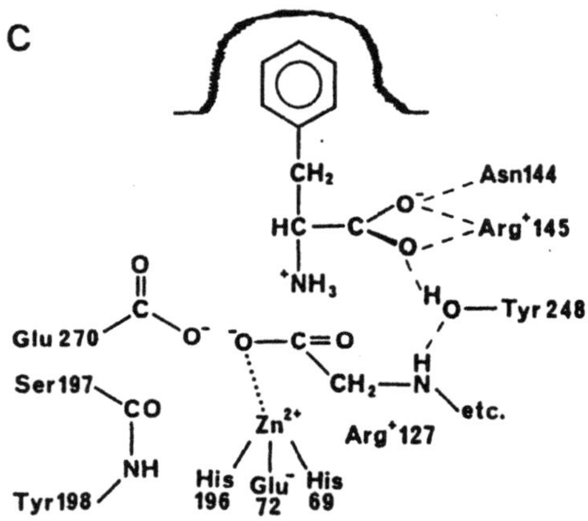

Fig.11. Proposed mechanism for CPA-catalyzed proteolysis: (A) the Michaelis complex with the substrate carbonyl hydrogen bonded to Arg-127; (B) the zinc and Arg-127 stabilized tetrahedral intermediate; (C) product complex after an intervening proton transfer

Summary

We propose that a new classification, distinct from that of "transition state analog", be established for those enzyme inhibitors which undergo the same reversible reactivity, but not the completion, which actual substrates undergo when bound to an enzyme active site. These "reaction coordinate analogs" do not merely imitate the structure of one point along the proteolytic reaction coordinate, but they undergo a reversible chemical transformation along the same reaction coordinate which defines catalysis (Fig.1). Reaction coordinate analogs, although they initially resemble unreacted substrate, are more appropriate for structural mechanistic study since they undergo a mechanistically pertinent chemical reaction. This is quite contrary to the chemical reactivity of so-called mechanism-based inhibitors, which undergo transformations which are not necessarily along the catalytic reaction coordinate. Another advantage to the study of reaction coordinate analogs is the opportunity for an enzyme to select the highest affinity form of the chemically transformed inhibitor. For example, a proteolytic enzyme's selection of a structure which involves a covalent bond with the enzyme or the addition of a water molecule to the inhibitor would provide structural support for nucleophilic or promoted-water hydrolytic pathways, respectively. The examples of reaction coordinate analogs described bound to the CPA active site indicate that the enzyme may favor a promoted-water hydrolytic pathway.

To visually summarize the structural-mechanistic study of CPA described herein, we provide Fig.11, which incorporates all mechanistic features which have been discussed. A noteworthy feature includes the participation of Arg-127 throughout the catalytic cycle. Both the zinc ion and Arg-127 can help to stabilize the negative charge of the oxyanion intermediate, as well as that of product carboxylate. The role of Glu-270 is that of general base/general acid, and all other residues depicted in Fig.11 elegantly determine the features of substrate specificity. Consistent with the results of Gardell and colleagues (Gardell et al. 1985) that the Phe-248 mutant is almost fully functional, we assign no obligatory catalytic role to Tyr-248 of the wild-type enzyme. We note that Rees and Lipscomb (1982) first considered Glu-270 as an intermediate proton donor in catalysis, and that Monzingo and Matthews (1984) recently elaborated on this catalytic role. Finally, we emphasize that CPA, and possibly many other zinc proteases, may possess not one, but two, catalytically important electrophiles in their active sites: in CPA these are divalent zinc ion and Arg-127. Further structural and mechanistic study may allow for elaboration on the catalytic contributions of these two active site electrophiles.

Acknowledgments. We thank the National Institutes of Health for Grant GM 06920 in support of this research. Additionally, D.W.C. thanks AT&T Bell Laboratories for a doctoral fellowship during his tenure of graduate study at Harvard University, during which time much of the work herein described was completed.

References

Auld DS, Holmquist B (1974) Biochemistry 13:4355-4361
Bradshaw RA, Ericsson LH, Walsh KA, Neurath H (1969) Proc Natl Acad Sci USA 63: 1389-1394
Burgi HB, Dunitz JD (1983) Acc Chem Res 16:153-161
Christianson DW, Lipscomb WN (1985) Proc Natl Acad Sci USA 82:6840-6844
Christianson DW, Lipscomb WN (1986a) Proc Natl Acad Sci USA 83:7658-7572
Christianson DW, Lipscomb WN (1986b) J Am Chem Soc 108:4998-5003
Christianson DW, Lipscomb WN (1986c) J Am Chem Soc 108:545-546
Christianson DW, Lipscomb WN (1987) J Am Chem Soc 109:5536-5538
Christianson DW, Lipscomb WN (1988) J Am Chem Soc 110:5560-5565
Christianson DW, Kuo LC, Lipscomb WN (1985) J Am Chem Soc 107:8281-8283
Christianson DW, David PR, Lipscomb WN (1987) Proc Natl Acad Sci USA 84:1512-1515
Galardy RE, Kortylewicz ZP (1984) Biochemistry 23:2083-2087
Gardell SJ, Craik CS, Hilvert D, Urdea MS, Rutter WJ (1985) Nature 317:551-555
Gelb MH, Svaren JP, Abeles RH (1985) Biochemistry 24:1813-1817
Grobelny D, Goli UB, Galardy RE (1985) Biochemistry 24:7612-7617
Hartsuck JA, Lipscomb WN (1971) In: Boyer P (ed) The enzymes, 3rd edn. Academic Press, New York, pp 1-56
Izumiya N, Uchio H (1959) J Biochem (Tokyo) 46:235-245
Jacobsen NE, Bartlett PA (1981) J Am Chem Soc 103:654-657
Lewis CA, Wolfenden R (1977) Biochemistry 16:4886-4890
Lienhard GE (1972) Annu Rep Med Chem 7:249-258
Lienhard GE (1973) Science 180:149-154
Lipscomb WN (1980) Proc Natl Acad Sci USA 77:3875-3878
Lipscomb WN (1982) Acc Chem Res 15:232-238

Lipscomb WN (1983) Annu Rev Biochem 52:17-34
Makinen WN, Kuo LC, Dymowski JJ, Jaffer S (1979) J Biol Chem 254:356-366
Monzingo AF, Matthews BW (1984) Biochemistry 23:5724-5729
Oren D, Ewenson A, Gilon C, Shoham G (1988) Biochem Biophys Res Commun (in press)
Pauling L (1946) Chem Eng News 24:1375-1377
Pocker Y, Sarkanen S (1978) Adv Enzymol 47:149-274
Quiocho FA, Lipscomb WN (1971) Adv Protein Chem 25:1-78
Rees DC, Lipscomb WN (1982) J Mol Biol 160:475-498
Rees DC, Lipscomb WN (1983) Proc Natl Acad Sci USA 80:7151-7154
Sander ME, Witzel H (1985) Biochem Biophys Res Commun 132:681-687
Shoham G, Christianson DW, Oren DA (1988) Proc Natl Acad Sci USA 85:684-688
Sugimoto T, Kaiser ET (1978) J Am Chem Soc 100:7750-7751
Sugimoto T, Kaiser ET (1979) J Am Chem Soc 101:3946-3951
Suh J, Cho W, Chung S (1985) J Am Chem Soc 107:4530-4535
Vallee BL, Galdes A (1984) Adv Enzymol 56:283-430
Vallee BL, Galdes A, Auld DS, Riordan JF (1983) In: Spiro TG (ed) Metal ions in biology, vol 5. Wiley, New York, pp 25-75
Van Wart HE, Vallee BL (1978) Biochemistry 17:3385-3394
Wolfenden R (1972) Acc Chem Res 5:10-18
Wolfenden R (1976) Annu Rev Biophys Bioeng 5:271-306
Yanari S, Mitz MA (1956) J Am Chem Soc 79:1154-1158

The Specific Interaction of Human Leukocyte Elastase with Various Protein Inhibitors

W. Bode [1]

Introduction

Human leukocyte elastase (HLE) is a major protein of the azurophilic granules of human polymorphonuclear granulocytes (Fig.1). It is believed to be normally involved in the intracellular degradation of cellular and phagocytosed proteins and bacteria. Upon stimulation it is, however, released into extracellular space: into the plasma, where it is normally blocked by the α_1-proteinase inhibitor (α_1PI), and into mucous secretions, where it can be inhibited by the mucous proteinase inhibitor (MPI).

In cases of inherited or acquired inhibitor deficiency (both inhibitors are, for example, sensitive to oxidation) or in other cases of excessive release (e.g., inflammation, septicemia) the released HLE is no longer properly balanced, so that it can degrade connective tissue proteins such as elastin, collagens, or proteoglycans, thus giving rise to the development of many organic dysfunctions (edema, shock lung, pulmonary emphysema) and cartilage erosion (Fig.1).

One promising approach to limit or even to prevent these damages is to increase the anti-elastase level by supplementing with appropriate inhibitors, namely with natural and recombinant inhibitors (for example, α_1-PI, MPI, eglin, variants based on the small Kunitz and Kazal-type inhibitors) and with small synthetic inhibitors. To understand their interaction with HLE and to rationalize the design and engineering of better inhibitors, we have undertaken the crystal structure analyses of HLE and of some potent elastase inhibitors.

Human Leukocyte Elastase (HLE)

We have been able to crystallize HLE (kindly provided by Drs. Neumann and Lang, Merck, Darmstadt) in complexes formed with the Turkey ovomucoid inhibitor third domain (OMTKY3, a kind gift of Dr. M. Laskowski Jr., Purdue University) and with a valyl chloromethylketone (MPM, kindly provided by Dr. J. Powers, Georgia Institute of Technology). The HLE-OMTKY3 complex has been solved by multiple isomorphous replacement techniques (Bode et al. 1986a) and has been completely refined to a crystallographic R-value of 0.166 using data to 1.8 Å resolution (Bode in preparation). The HLE-MPM complex was solved by Patterson search techniques and completely refined to an R-value of 0.145 with data to 2.3 Å (Wei et al. 1988). The exact polypeptide sequence of HLE was established by combined efforts using crystallographic and peptide sequencing methods (Sinha et al. 1987).

[1] Max-Planck-Institut für Biochemie, D-8033 Martinsried, FRG

39. Colloquium Mosbach 1988
Protein Structure and Protein Engineering

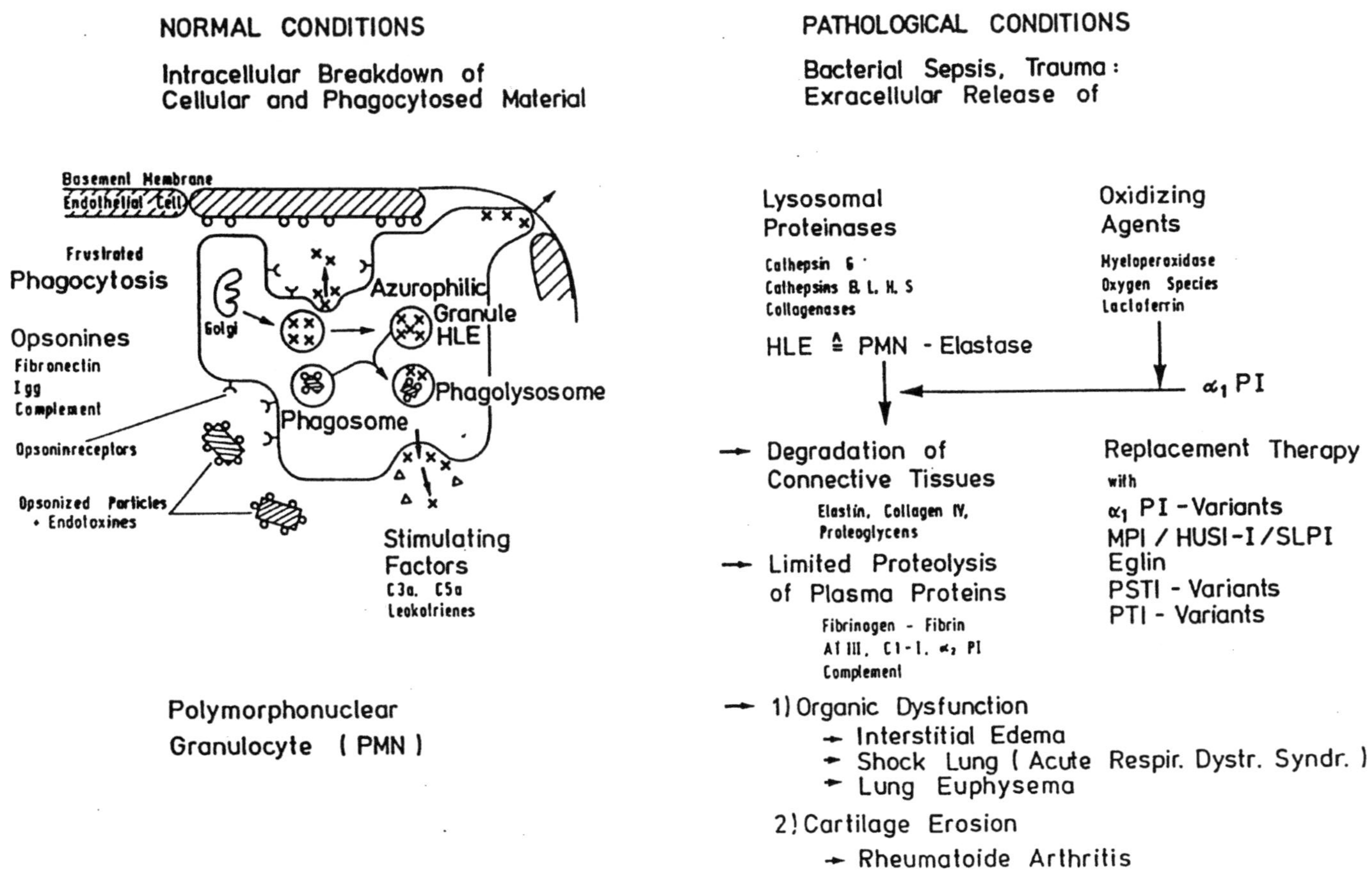

Fig. 1. Involvement of HLE in protein turnover, defense and development of organic disfunctions

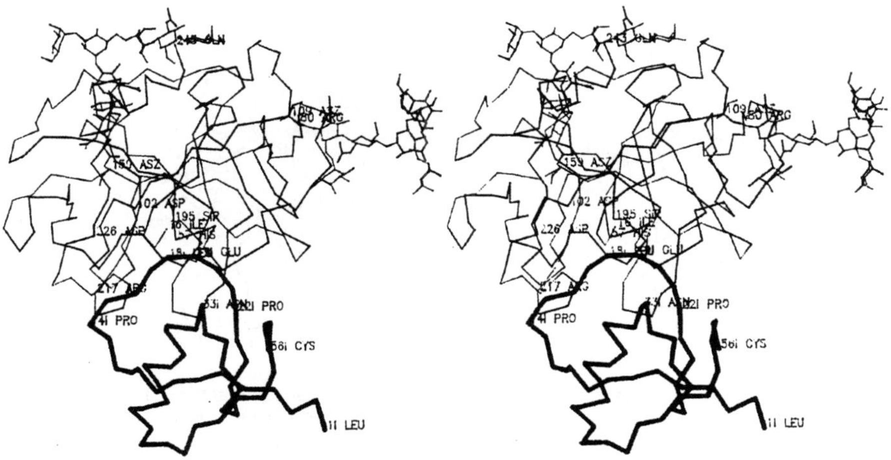

Fig. 2. α-carbon stereoplot of the molecular complex formed between HLE (thin connections) and OMTKY3 (bold connections). Both sugar chains and the linking asparagines are shown with all nonhydrogen atoms

Figure 2 shows an α-carbon drawing of the HLE-OMTKY3 complex. The main chain of HLE, comprising 218 amino acid residues, can be readily traced from its amino-terminal residue Ile16 (chymotrypsinogen numbering) to its carboxy-terminal residue Gln243. Due to its sequence and polypeptide fold, HLE is a typical vertebrate serine proteinase. Of the related enzymes for which tertiary structures have been evaluated, HLE is topologically most equivalent with porcine pancreatic elastase (PPE), followed (in that order) by rat mast cell protease II, porcine pancreatic kallikrein, rat tonin, and bovine α-chymotrypsin. Also, according to sequence alignment, HLE is most similar to PPE (with 40% identical residues). This similarity is further emphasized by four common disulfide bridges.

However, at some sites, all of which are situated on the molecular surface, the HLE chain deviates from the related proteinases (Bode et al. 1986a). The largest deviation occurs around the "methionine loop". There is no indication of an "intermediate helix" which is normally observed in the vertebrate proteinases around Cys168. The loop segment bewteen this Cys168 and Cys182 is of much smaller size and quite different conformation.

HLE is a glycoprotein; two of its three Asn-based glycosylation consensus sequences, at Asn109 and Asn159, are glycosylated (Bode et al. 1986a; Sinha et al. 1987). At both sites only the first three to four sugars (in each case an N-glycosidically attached N-acetyl-glucosamine, α1,6-linked to a terminal α-L-fucopyranose and β1,4-linked to a further N-acetyl-glucosamine, which in turn is connected to another sugar, probably a branching mannose) are defined by proper electron density. Most of the noncovalent contacts to the protein moiety are made by both fucose rings. Both sugar chains, parts of which are probably of the complex type (Sinha et al. 1987), extend out into solution and protrude away from the substrate binding site (Fig.2); consequently, they should not interfere with binding of substrates and inhibitors.

Since 19 arginines, but only nine acidic residues are present, HLE is an extremely basic protein (Ip about 11). All of these arginine residues

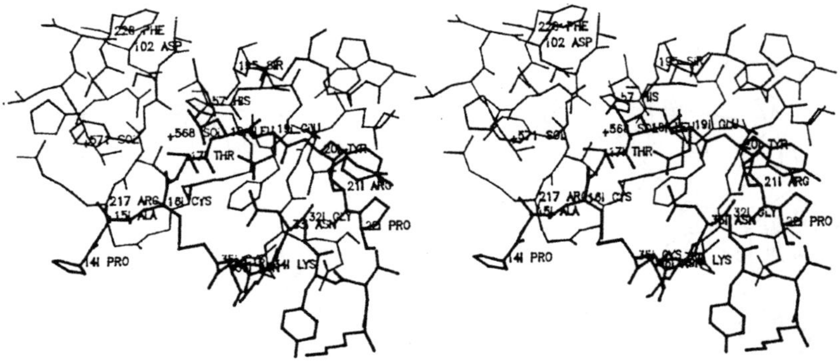

Fig. 3. Stereo view of the HLE-OMTKY3 interface, with contacting HLE residues (thin connections) and OMTKY3 segments Pro14I - Pro22I and Lys29I - Asn36I (bold connections). Orientation as in Fig.2

(with the exception of Arg80) are arranged in a horseshoe-like manner around the binding site, some of them forming clusters of arginines. Their surface arrangement explains the preferred binding of linear sulfated polysugars by HLE; the inhibitory effect of these compounds might be due to interaction with Arg217 (which is localized closest to the active site residues), either through sterical hindrance or through conformational changes.

HLE-segment 70-80 is spatially similar to the "calcium-loop" in trypsin and PPE; in HLE, however, the terminal guanidyl group of Arg80, which replaces Glu80 in the calcium stabilized enzymes, occupies a site equivalent to the calcium ion in those enzymes. Thus HLE carries its "own" stabilizing cation and does not seem (in agreement with solution experiments) to depend on external calcium.

The active site residues (Ser195, His57, Asp102, Ser214) are similarly positioned in HLE as in the other vertebrate serine proteinases. The substrate binding cleft more distant from these residues shows, however, considerable differences. Remarkable amino acid replacements within the three HLE strands lining the specificity pocket (compared with PPE) occur at positions 189 (Ser(PPE) → Gly(HLE), 190 (Gly → Val), 192 (Gln → Phe), 213 (Thr → Ala), 217 (Ser → Arg, with an extra residue in PPE), 226 (Thr → Asp) and 227 (Val → Ala). Identical (and possibly typical for elastases) are Phe215, Val216, and Phe228. Unique to HLE is the occurrence of an acidic residue (Asp226) just beyond the back of the S1-pocket, whose side chain is surrounded by a few internally bound solvent molecules; the significance of this particular structural element is yet still unclear.

HLE-OMTKY3 Interaction

OMTKY3 (Laskowski et al. 1987) exhibits the characteristic ovomucoid structure mainly consisting of a three-stranded β-pleated sheet, an α-helical segment and an extended proteinase binding loop which contains the reactive site Leu18I-Glu19I. The inhibitor is in direct contact with the binding site of the cognate enzyme (Fig.3) through eight residues of this latter "primary binding segment" (Pro14I, P5, to Arg21I,

P3') and through additional three (to five) residues (Gly32I, Asn33I, and Asn36I) of a "secondary binding" segment. The molecular surfaces removed from access to bulk water upon complex formation are nevertheless relatively small (600 Å^2 for each component). The reactive site loop spanned between Pro14I and Pro22I (Fig.3) in an overall extended conformation neatly fits to the enzyme's binding cleft, with the exposed side chains of the residues which flank the reactive site bond alternatively protruding into opposite directions.

As known from the analysis of isolated ovomucoid inhibitors, this conformation (complementary to the binding site of the cognate enzyme) is an intrinsic property of the inhibitor itself. Complex formation occurs without major conformational changes, but under rigidification of the inhibitor's binding segment.

The most intimate contact with enzyme groups are made by P1-residue Leu18I of OMTKY3. Its side chain is in the energetically most favorable (g^+-t)-conformation and extends into the specificity pocket of HLE. This pocket is constricted towards its bottom by the hydrophobic residues Val190, Phe192, Ala213, Val216 and Phe228 and contains no solvent molecule. In spite of the partially differing amino acids, the hydrophobic nature is similar to the PPE pocket, and the accessible space is apparently only marginally larger than that in PPE, thus explaining the preference of HLE for slightly larger, but still medium-sized hydrophobic side chains at P1. Modeling and energy minimization studies indicate that (in agreement with experimental results) elongated aliphatic residues should fit even better into the S1-subsite of HLE.

Methionine can be easily modeled in a bent conformation into this pocket; the extra oxygen of a methionine sulfoxide should, however, create severe sterical hindrance, thus explaining the observed dramatic affinity decrease of oxidized inhibitors with methionine at P1 (or the decreased activity of equivalent substrates). There is, of course, enough space to accommodate smaller side chains. As shown by a comparison with the HLE-MPM structure, a P1-valine, with a β-branched bulky side chain, requires some slight tilting at its main chain, in order to avoid unfavorable contacts; the valine side chain is arranged in a less frequently observed (g^-)-conformation, and the pocket adapts slightly to the reduced size of the side chain.

Also the other contacting residues of the primary binding segment fit well to the enzyme's binding cleft. According to the number of intermolecular contacts, P2' and P2-residues Tyr20I and Thr17I rank next to P1-Leu18I. The complementary subsites are not as srongly invaginated as the S1-specificity pocket, but still significantly concaved and equipped to explain the preference of HLE for these inhibitor residues. All other residues of the inhibitor are, in addition to their interaction with enzyme groups, to a high degree exposed to bulk water molecules, thus making an estimation of their contribution to binding difficult.

We have (in collaboration with M. Laskowski) replaced in the HLE-OMTKY3 model several of the contacting ovomucoid residues and tried to predict the effects of such replacements. The third domains from over 100 avian species have been characterized yielding 65 different sequence variants (Laskowski et al. 1987) with exchanges occurring mainly in the contact region. Currently, M. Laskowski is determining binding constants for these variants towards HLE, in this way obtaining binding increments for different amino acids at the distinct inhibitor positions. Of particular importance for binding are the P2, P2', and P14'-residues. The results obtained to date are in relatively good agreement with the predictions derived from the model (M. Laskowski, personal communication). These data will allow prediction and design of HLE inhibitor variants of high affinity or extreme selectivity.

PSTI- and BPT-Based HLE-Inhibitors

Based on the human pancreatic secretory trypsin inhibitor, which exhibits a chain fold similar to the Kazal-type ovomucoid inhibitors, but is, due to a P1-lysyl residue, specific for trypsin-like enzymes, and led by the results obtained for the ovomucoid-HLE interactions, scientists from the GBF in Braunschweig and from the Bayer AG in Wuppertal have succeeded in preparing (by recombinant techniques) inhibitors with quite good association rate and binding constants towards HLE (Collins et al. 1987). Main features of these HLE-variants were leucine or valine as P1-residues and arginine as P3'-residue.

Scientists in Bielefeld (Tschesche et al. 1987) have used semi-synthetic procedures, and workers at the Bayer AG recombinant techniques to convert BPTI, an inhibitor with specificity towards trypsin, into HLE-specific variants. The first approach is particularly attractive due to the possibility to insert unusual amino acids or even nonamino acid residues. Interestingly, in contrast to ovomucoid inhibitors, valine at P1 makes these BPTI-based variants better HLE-inhibitors than leucine. BPTI possesses a binding segment of a geometry quite similar to that of the ovomucoid inhibitors between P3 and P4', but is otherwise completely unrelated in structure.

Subtilisin Carlsberg - Eglin c

The X-ray crystal structure of eglin c, an elastase inhibitor from medical leech, has been solved in a complex formed with subtilisin Carlsberg (Bode et al. 1986b,1987) and has been completely refined to an R-value of 0.178 with data to 1.2 Å resolution (Fig.4). Eglin c comprising in the intact form 70 amino acid residues consists of a four-stranded β-sheet with an α-helical segment and the proteinase binding loop fixed on opposite sides (Fig.5). Eglin c is completely differently folded compared with the small Kunitz and the Kazal-type inhibitors. It possesses, however, an extended binding loop of a conformation (and in part also sequence) very similar to that observed in the other HLE-specific inhibitors, which allows a favorable binding to subtilisin, but also (as shown experimentally and by model building) to HLE. Eglin lacks any disulfide bridge which could stabilize this binding loop around the reactive site bond Leu45I - Asp46I. Instead, two almost extended arginyl side chains (Arg51I and Arg53I) connect it through polar interactions with the hydrophobic core of the molecule and keep it in a conformation suitable for binding.

HLE-MPI Complex

For immunological reasons, the availability of protein inhibitors of human origin is desirable. The two inhibitors probably most responsible for security blockage in human beings are α_1-PI and MPI. The latter inhibitor, consisting of 107 amino acid residues, is found in various

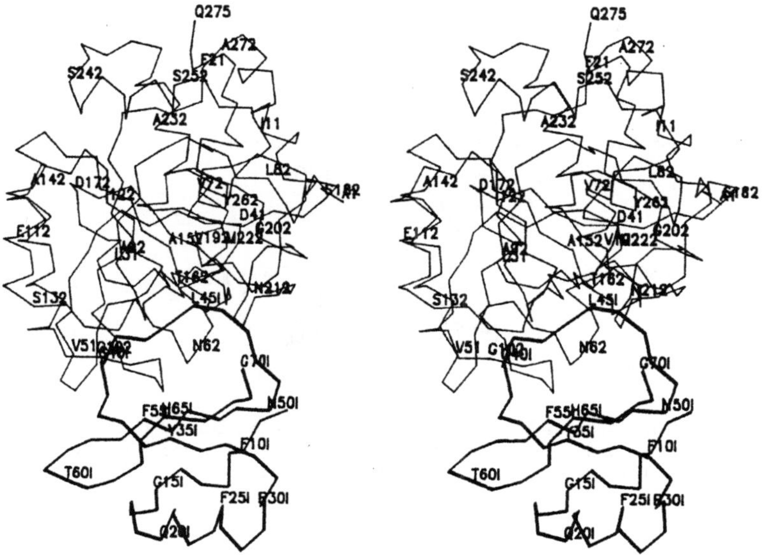

Fig. 4. α-carbon stereoplot (Bode et al. 1987) of the molecular complex formed between subtilisin Carlsberg (thin connections) and eglin c (bold connections)

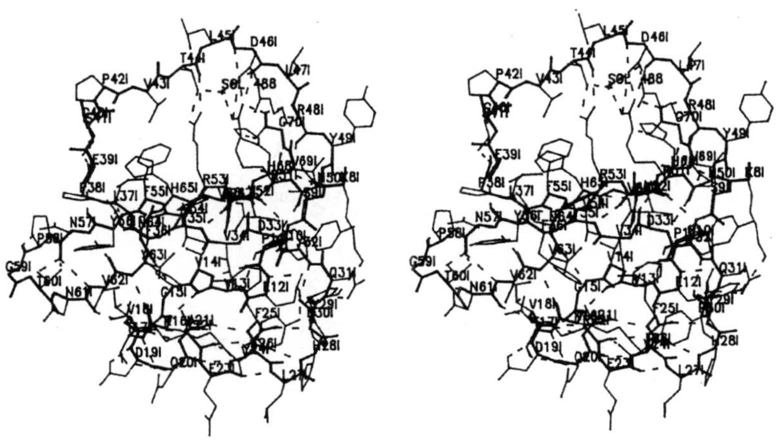

Fig. 5. α-carbon stereoplot (Bode et al. 1986b) of eglin c. Probable intramolecular hydrogen bonds are shown by dashed lines

human mucous fluids, where it occurs in multiple forms due to varying proteolytic cleavage events. Depending on the tissue of origin these inhibitors have been called human seminal plasma inhibitor I (HUSI-I), bronchial mucous inhibitor (BMI), secretory leukozyte proteinase inhibitor (SLPI), etc. Due to the common origin from mucous fluids, H. Fritz (University of Munich) has suggested the term mucous proteinase inhibitor (MPI). In collaboration with Dr. M. Grütter (Ciba-Geigy,

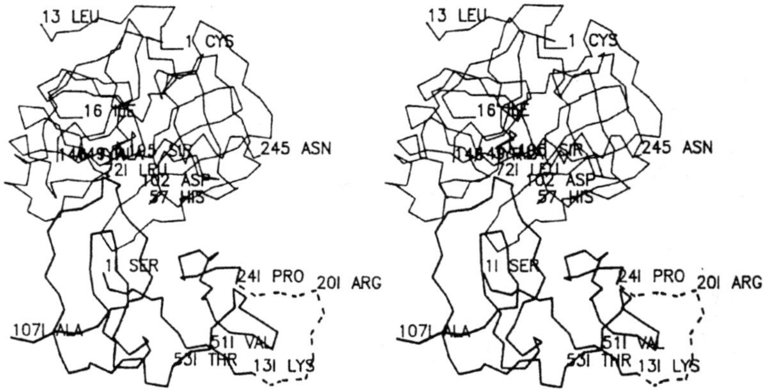

Fig. 6. α-carbon stereoplot (Grütter et al. 1988) of the complex formed between bovine α-chymotrypsin (thin connections) and human MPI/SLPI (bold connections). The disordered MPI segment Lys13I - Pro24I is given in dashed lines

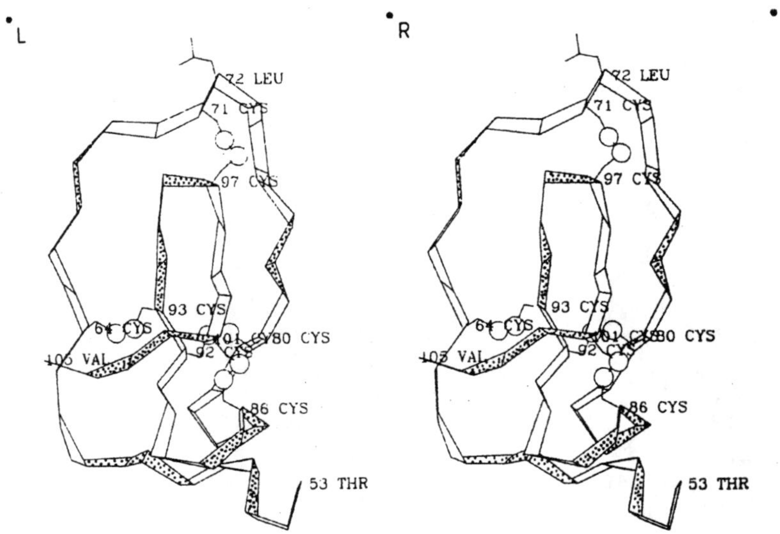

Fig. 7. Ribbon diagram stereoplot (Grütter et al. 1988) of the polypeptide chain of the second MPI/SLPI domain

Basel), we have been able to crystallize a bovine α-chymotrypsin complex of a recombinant human SLPI/MPI (prepared by Synergen, Boulder, Co) and solved its molecular structure (Fig.6), applying Patterson search techniques (Grütter et al. 1988).

MPI/SLPI has a boomerang-like shape with wings comprising two well-separated domains of similar architecture. These domains are made up of the amino-terminal and the carboxy-terminal halves of the polypeptide chain, respectively, and represent the anti-tryptic and the anti-elastolytic part of the inhibitor. In each domain the polypeptide chain is arranged like a stretched spiral (Fig.7). Two internal strands form a regular β-hairpin loop which is accompanied by two external strands linked by the proteinase binding segment. The polypeptide segment of each domain is interconnected by four disulfide bridges with a connectivity pattern hitherto unobserved.

The binding loop of the first domain is disordered due to proteolytic cleavage occurring in the course of crystallization. The anti-tryptic characteristics of this domain suggest Arg20I-Tyr21I as the reactive site bond, which makes it (with respect to the preceding cystine bridge) similar to Kazal-type inhibitors.

The reactive site loop of the second domain is of similar conformation as the binding segments of the other "small" inhibitors. Model building experiments show that MPI/SLPI also fits to the binding site of HLE without requiring large conformational changes. Due to its smaller specificity pocket, the fit of the P1-Leu72I to HLE should be even better than to chymotrypsin. Eight residues of this loop, two of the adjacent hairpin loop, the carboxy-terminal segment, and Trp30I of the first domain are in direct contact with the cognate enzyme. Surprisingly, and in contrast to the first domain, the reactive site bond was found between Leu72I and Met73I, that is only one residue behind the cross-linking cystine (Cys71I), which makes this domain (in this respect) look more like a BPTI-type inhibitor. The oxidation sensitivity of MPI/SLPI and the analogy to the binding segment of the first domain had made the scientists involved in sequencing assume that Met73I might be the P1-residue. The close packing of the side chains of two methionine residues (Met74I and Met96I) against α-chymotrypsin could explain the oxidation sensitivity of the anti-elastase property of MPI/SLPI.

The only few interdomainal contacts and the quite accurate folding agreement between both domains (the r.m.s. deviation of all α-carbon atoms of both domains is 0.7 Å) suggest an almost independent folding and stabilization of both domains. A striking property of the second domain is the significant exposure of the quite hydrophobic binding segment and the simultaneous burying of at least two charged side chain groups, thus apparently contradicting the generally accepted dogma about amino acid distribution in protein structures.

Human α_1-PI

The inhibitor responsible for elastase inhibition in our plasma is α_1-PI. Some years ago we have been able to grow crystals suitable for X-ray diffraction analysis not of intact, but of modified human α_1-PI, which had specifically been cleaved at the reactive site bound Met358-Ser359 by virtue of chymotrypsinogen (Löbermann et al. 1982). Its structure was solved in two different crystal forms at 3.0 Å resolution using multiple isomorphous replacement techniques and molecular averaging methods (Löbermann et al. 1984) and has recently been crystallographically refined.

The modified inhibitor mainly consists (besides several α-helices primarily made up of the amino-terminal part of the polypeptide chain) of three distinct β-pleated sheet elements. Surprisingly, the newly formed chain ends of this nicked species (Met358 and Ser359) are arranged on opposite sides of the ellipsoidal molecule (almost 70 Å apart from one another), and their peptide strands are integral members of different β-sheets. Before the cleavage, both ends must have been covalently connected; possibly in a loop-like fashion similar to that found in eglin, resulting in an interaction with the cognate enzyme similar to that observed for the "small" inhibitors. Human α_1-PI possesses a methionine at the P1-position which is easily oxidized, e.g., upon smoking or other stimulating events. The oxidized inhibitor has a lower affinity to HLE (which is plausible from the knowledge of the HLE-structure) and reacts much more slowly, so that it can no longer compete with elastin for binding of HLE. A few α_1-PI variants have been prepared by recombinant methods (mainly by replacing its P1-methionine by an oxidation resistant valine) yielding quite efficient HLE inhibitors.

Conclusion

There are several types of serine proteinase inhibitors of quite unrelated peptide fold (Laskowski and Kato 1980), all of which possess an extended loop segment of a common, unique conformation, which allows an unconstrained binding to their cognate enzymes in a manner similar to normal peptide substrates. A few residues, in particular the P1-residue, determine the specificity of the inhibitor variants towards distinct proteinases as HLE. Their side chains are protruding from the exposed binding loop and can thus be exchanged without major effects on the inhibitor architecture. Single residue substitutions can, however, lead to completely changed specificity; the remarkable simplicity of this switching mechanism might have caused nature to develop this specific type of serine proteinase inhibition and to apply it to such a large extent.

When the principal structures involved are known, new inhibitors with changed or improved specificity can be designed and engineered with modern biochemical methods. Due to their specific structure and their substrate-like interaction with cognate enzymes, serine proteinase inhibitors seem to be particularly suited to such manipulations. As we see now for the inhibition of cysteine proteinases (Bode et al. 1988), the specificity of protein inhibitors directed towards the other proteinase families might not be manipulated in a similarly simple way.

References

Bode W, Wei AZ, Huber R, Meyer E, Travis J, Neumann S (1986a) EMBO J 5:2453-2458

Bode W, Papamokos E, Musil D, Seemüller V, Fritz H (1986b) EMBO J 5:813-818

Bode W, Papamokos E, Musil D (1987) Eur J Biochem 166:673-692

Bode W, Engh R, Musil D, Thiele W, Huber R, Karshikov A, Brzin J, Kos J, Turk V (1988) EMBO J 7:2593-2599

Collins J, Blöcker H, Maywald F, Ebbers J, Gross G, Meyerhans A, Schwellnus K, Böldicke T, Böcher M, Szardenings M, Frank R, Fritz H (1987) Febs Sattelite Meeting on Proteinase Inhibitors and Biological Control, Ljubljana/BRDO, July 4-7, 1987 (organized by H. Fritz and V. Turk)

Grütter MG, Fendrich G, Huber R, Bode W (1988) EMBO J 7:345-351

Laskowski M Jr, Kato I (1980) Annu Rev Biochem 49:593-626

Laskowski M Jr, Kato I, Ardelt W, Cook J, Denton A, Empie MW, Kohr WJ, Park SJ, Parks K, Schatzley BL, Schoenberger OL, Tashiro M, Vichot G, Whatley HE, Wieczorek A, Wieczorek M (1987) Biochemistry 26:202-221

Löbermann H, Lottspeich F, Bode W, Huber R (1982) Hoppe-Seyler's Z Physiol Chem 363: 1377-1388
Löbermann H, Tokuoka R, Deisenhofer J, Huber R (1984) J Mol Biol 177:531-556
Sinha S, Watorek W, Karr S, Giles J, Bode W, Travis J (1987) Proc Natl Acad Sci USA 82:2228-2232
Tschesche H, Beckmann J, Mehlich A, Schnabel E, Truscheit E, Wenzel H (1987) Biochim Biophys Acta 913:97-101
Wei AZ, Mayr I, Bode W (1988) FEBS Letters 234:367-373

The Interplay Between Enzyme Mechanism, Protein Structure, and Inhibitor and Catalyst Design

P. A. Bartlett [1]

Introduction

As suggested by the tetrahedron in Fig.1, there are obvious relationships between enzyme mechanism, protein structure, and the design of enzyme inhibitors and mimics. Some of the edges of this tetrahedron are stronger than others, and between some of the vertices information flows in one direction more than the other. However, the accelerating progress in protein structure determination and the recently made connection between inhibitor and catalyst design are beginning to address this imbalance. In this presentation, a number of facets of our research will be discussed in which the relationships depicted in Fig.1 are important.

Strategies for Inhibitor Design: Mechanism-Derived

The most effective strategies for the design of enzyme inhibitors involve the synthesis of "suicide inhibitors" (Abeles 1980; Walsh 1982) or of transition state analogs (Wolfenden 1976; Stark and Bartlett 1984). In suicide inhibition, the catalytic machinery of the enzyme is used to convert an otherwise innocuous molecule into one that will inactivate the enzyme. A transition state analog is intended to resemble the high energy species along the reaction pathway and take advantage of the special affinity which the enzyme has for this structure In both approaches the important information that goes into designing

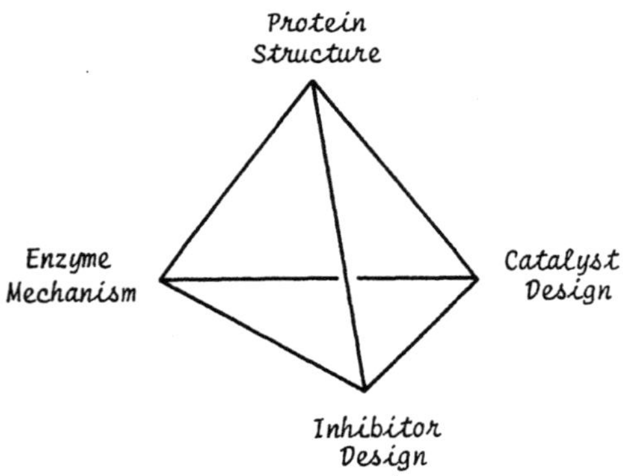

Fig. 1

[1]Department of Chemistry, University of California, Berkeley, California 94720, USA

39. Colloquium Mosbach 1988
Protein Structure and Protein Engineering

the inhibitor comes from a knowledge of the mechanism of the enzymatic transformation. Both types of inhibitors should therefore be considered "mechanism-derived".[2]

Phosphonates as Transition State Analogs of Thermolysin

An illustration of the flow of information is provided by the evolving story of phosphorus-containing peptide derivatives as inhibitors of the zinc peptidase thermolysin. By showing that the binding affinities of a series of such inhibitors (**1**) correlate with the inverse second order rate constants for enzymatic turnover of the analogous substrates, we were able to show that these analogs are indeed transition state analogs (Bartlett and Marlowe 1983).

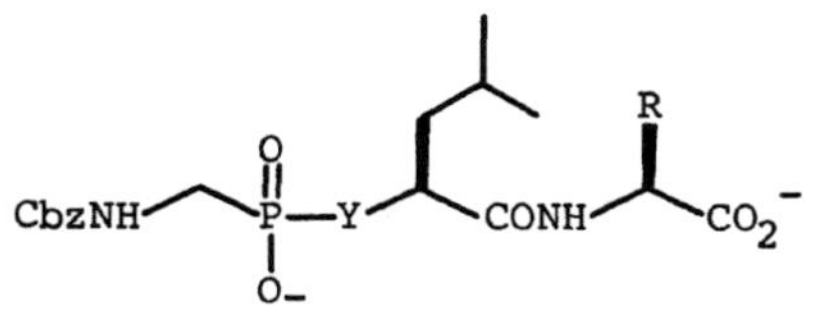

1: Y = NH

2: Y = O

In turn, the crystal structures of several such inhibitors bound to the active site of thermolysin (Tronrud et al. 1987; Holden et al. 1987) reveal a number of specific interactions between the enzyme and the inhibitors which, as models of the transition state structure, shed light on some of the details of the enzyme mechanism. For example, the relationship between the catalytic carboxylate of glutamate-143 and the inhibitor coordinated to the zinc ion suggests a specific role for this residue in shuttling protons from the attacking water molecule to the departing amine (Hangauer et al. 1984).

A further specific interaction was revealed both crystallographically and through a comparison of the phosphonamidates **1** and phosphonate ester analogs **2**. The latter compounds bind uniformly 840-fold more weakly than the former, representing a difference in binding energy of 4.0 kcal/mol (Bartlett and Marlowe 1987a). That this difference in affinity can be ascribed to the loss of a hydrogen-bonding interaction in going from the amidates to the esters is underscored by the crystallographic studies, which revealed virtually identical structures for appropriately paired complexes (Tronrud et al. 1987; Matthews, personal communication).

Phosphonates as Transition State Analogs of Carboxypeptidase A

Although there are strong similarities in active site structure between thermolysin and carboxypeptidase A, there has seemingly been more controversy surrounding the mechanism of the latter enzyme than the former (inter alia: Christianson and Lipscomb 1987; Sander and Witzel 1985).

[2]It is recognized that this characterization of "mechanism-derived" inhibitors may cause problems with the appellation "mechanism-based", which has often been used specifically for suicide inhibitors; it is hoped that the latter, more ambiguous, term will fall into disuse.

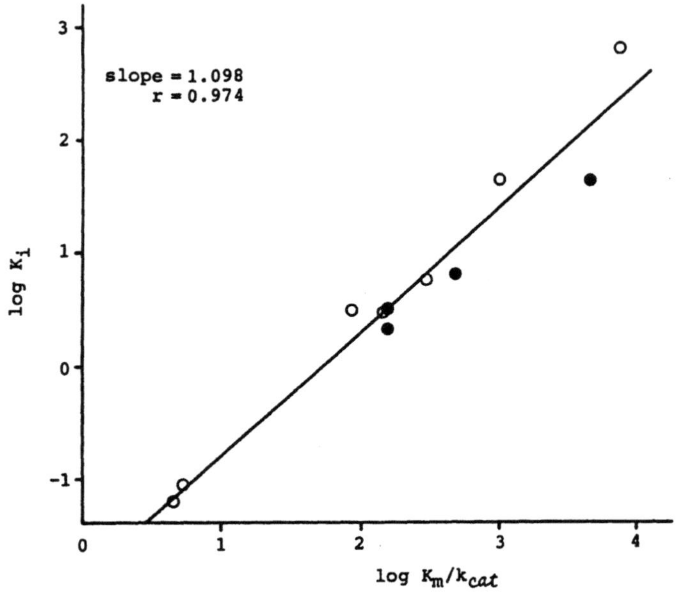

Fig. 2. Comparison of K_i values for the tripeptide (o, 3a) and tetrapeptide inhibitors (•, 3b) with the K_m/k_{cat} values for the corresponding substrates (data from Abramowitz-Kurn and Schechter 1974); K_i values determined at pH 7.5, 25°C

Indeed, the question of glutamate-as-base or glutamate-as-nucleophile remains an active issue. If the mechanism of carboxypeptidase A is the same as that of thermolysin, then phosphonate derivatives should be transition state analog inhibitors of the former enzyme as well. Such, in fact, is the case: for a series of 11 tri- and tetrapeptide analogs of the form 3, a strong correlation between inhibitor K_i and substrate K_m/k_{cat} values (Abramowitz-Kurn and Schechter 1974) can be demonstrated (Fig.2). Since these inhibitors cannot mimic the tetrahedral intermediate(s) involved in the glutamate-as-nucleophile mechanism, such a correlation is further evidence in support of the alternative in which glutamate serves as a base, in analogy with thermolysin.

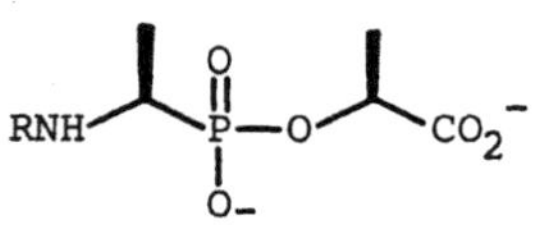

3a: R = Cbz-Phe, Cbz-Ala, Cbz-(D)Ala, Cbz-Gly, Phenylpropanoyl-Ala, Ac-Ala, Ac-(D)Ala

3b: R = Cbz-Phe-Ala, Cbz-Leu-Ala, Cbz-Ala-Ala, Cbz-Ala-(D)Ala

Carboxypeptidase A is selective for cleavage of C-terminal aromatic amino acids from a polypeptide chain, yet the correlation of Fig.2 was carried out with alanine at this position. The reason for this is two-fold. First, as described below, the analogous inhibitors containing

Table 1. Inhibition of carboxypeptidase A by phosphonate tripeptide analogs

Structure[a]	K_i (pM)[b]
Cbz-Ala-D-Ala^P-Phe	93,000
Cbz-Ala-Gly^P-(O)-Phe	710
Cbz-Ala-Ala^P-Phe	32
Cbz-Ala-Phe^P-(O)-Phe	4
Cbz-Ala-Ala^P-(O)-Phe	3
Cbz-Phe-Ala^P-(O)-Phe	1

[a]Ala^P represents the phosphonic acid analog of alanine; -(O)-Phe represents L-α-phenyllactate, the hydroxyl analog of phenylalanine. [b]Determined at 25°C in 0.5 M NaCl, 50 mM Tris-HCl at pH 7.5.

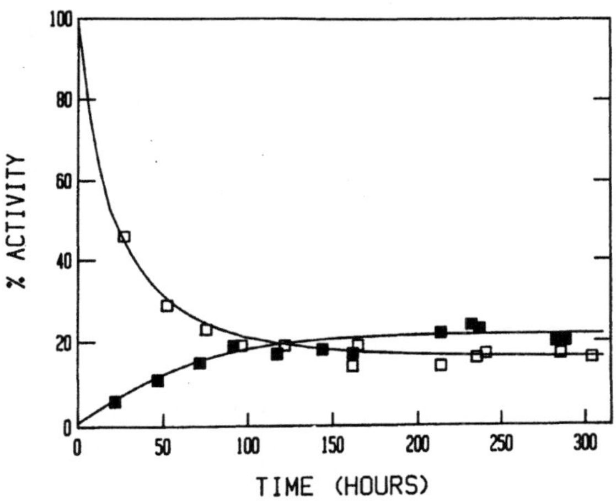

Fig. 3. Determination of K_i for inhibition of carboxypeptidase A by Cbz-Ala-Ala^P-(O)-Phe. (■) decrease of enzymatic acticity with time on combination of 45.4 pM CPA and 49.5 pM inhibitor. (□) increase of enzymatic activity in a solution containing 45 pM EI complex and 0.4 pM free inhibitor. The *curves* represent simulations of the equilibration using the following parameters: On-experiment: $k_{on} = 2.5 \times 10^5$/M/s, $k_{off} = 7.0 \times 10^{-7}$/s, $K_i = 2.8$ pM; Off-experiment: $k_{on} = 2.46 \times 10^5$/M/s, $k_{off} = 8.1 \times 10^{-7}$/s, $K_i = 3.3$ pM

a phenylalanine mimic at this position (4) are bound so tightly that determination of their equilibrium K_i values is not straightforward. Second, as was encountered with analogs of rapidly hydrolyzed substrates of thermolysin (Bartlett and Marlowe 1987b), the correlation breaks down when the chemical step is no longer the rate-limiting one for substrate turnover.

As revealed by the list in Table 1, phosphonate analogs which do incorporate a phenylalanine mimic at the P_1' position are exceedingly tightly bound. They may be the most potent synthetic inhibitors yet reported for any peptidase. For those inhibitors with K_i values below 1 nM, steady state methods could not be used to determine the inhibition constants, and separate determinations of k_{off} and k_{on} and of equilibrium binding to the enzyme in the absence of substrate were required (Fig.3; note the time course of the experiment).

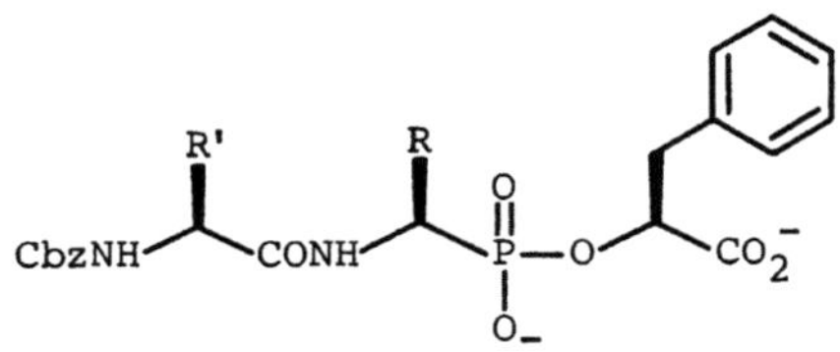

4

While a crystal structure is not yet available for a complex of carboxypeptidase A with one of the phosphonate tripeptide inhibitors, it is attractive to try to model the interaction, taking as a starting point the structure of the complex of CPA with the potato inhibitor (Rees and Lipscomb 1982). If the P_1 and P_2 residues of the phosphonate inhibitor are overlaid with the corresponding part of the backbone of the potato inhibitor, then not only can the C-terminal phenylalanine residue be fit appropriately into the S_1 binding site, but a relationship between the catalytic glutamate sidechain, the tetrahedral phosphonate, and the active site zinc ion can also be generated which is almost identical to that seen in the thermolysin complexes. It is also possible to rationalize the dramatically different effect of the O-for-NH sùbstition in CPA in comparison with thermolysin, since there is no hydrogen-bond acceptor in the vicinity of this moiety in CPA.

Catalyst Design

There has been a strong connection along the side of the tetrahedron of Fig.1 between enzyme mechanism and catalyst design, with considerable effort being devoted to juxtaposing side-chain residues and binding sites in imitation of the relationships found in enzyme active sites. What has recently been established is the connection between inhibitors and artificial catalysts, as a result of the demonstration that antibodies generated against transition state analogs can, in fact, act as catalysts. The concept is an elegantly simple one: if an enzyme accelerates its reaction by the complementarity of its active site to the transition state geometry, an immunoglobulin combining site which is elicited against a mimic of that structure should manifest similar characteristics.

Phosphonates as Templates for Catalytic Antibodies

This new connection on Fig.1 has also been established with tetrahedral phosphorus derivatives. Antibodies against a variety of phosphonate and phosphate derivatives have been shown to catalyze the hydrolysis of esters or other acyl transfer reactions (Tramontano et al. 1986; Pollack et al. 1986; Jacobs et al. 1987; Pollack and Schultz 1987; Napper et al. 1987; Schultz 1988). For example, monoclonal antibodies elicited against the cyclic phosphonate **5** catalyze the stereoselective lactonization of hydroxy-ester **6** (Napper et al. 1987). It is noteworthy that enzyme-catalyzed acyl transfers probably take advantage of acidic or basic residues in the active sites, in addition to control over the geometry of the substrates. That antibodies can catalyze this class of reaction suggests that the phosphonate is not only capable of inducing the appropriate geometry but also the necessary polarity in the combining site.

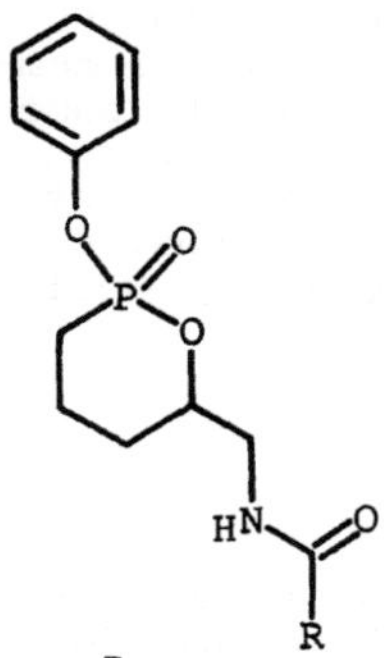

5

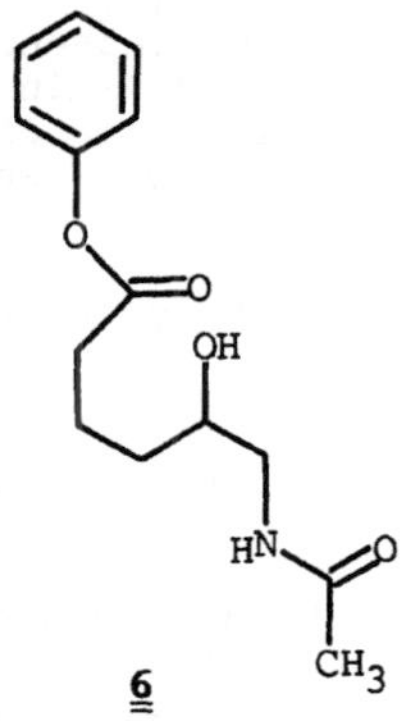

6

The Generation of a Chorismate Mutase Antibody

For the chorismate mutases there has been long-standing interest in the role which substrate conformational control plays in enzymatic rate acceleration (Guilford et al. 1987, and references therein). These enzymes catalyze the formal Claisen rearrangement of chorismic acid to prephenic acid [Eq.(1)], via a transition state conformation which resembles 7 (Sogo et al. 1984). Based on the premise that conformational control is an important factor in the enzymatic mechanism, bicyclic mimics of 7 have been synthesized as potential transition state analog inhibitors of the chorismate mutases. The most potent of these is the endo-diacid 8, which as the racemate inhibits the chorismate mutase/prephenate dehydrogenase from *E. coli* with a K_i value of 0.12 µM (Bartlett et al. 1988).

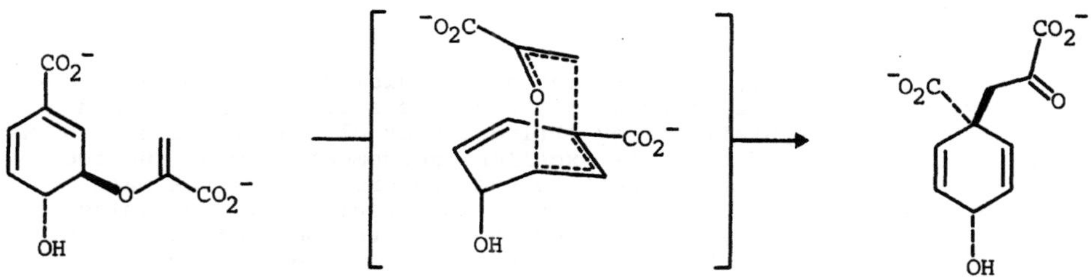

Chorismate 7 *Prephenate*

(1)

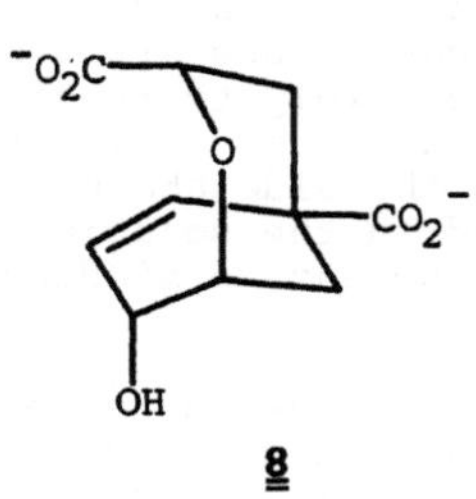

8

This inhibitor was derivatized through the hyeroxyl group, conjugated to a carrier protein, and used to induce immunoglobulins directed against it (Schultz et al. 1988). Of a number of monoclonals screened, one proved to be a chorismate mutase, showing saturation kinetics (K_m = 260 μM), inhibition by the hapten 8 (K_i = 9 μM), and acceleration of the rearrangement of chorismate to prephenate by a factor of 10,000 over the noncatalyzed rate (k_{cat} = 2.7/min at 10°C). This can be compared to the enzymatic rate acceleration of 3,000,000 under the same conditions. The fact that, in terms of transition state binding energy, the antibody demonstrates more than 60% of the rate acceleration of the enzyme-catalyzed transformation suggests that conformational control may well play a significant role in the latter process as well.

Strategies for Inhibitor Design: Structure-Derived

There remains another important edge to the tetrahedron of Fig.1. With the increasing availability of protein structures, both in solution as well as in the crystalline phase, and of computationally more sophisticated models for structures which have not been determined experimentally, an increasingly appealing strategy for the design of biologically active molecules is one that will be termed a "structure-derived" approach. This appellation is intended to contrast with the definition of "mechanism-derived" inhibitor design presented above, in that the structure of the enzyme active site, rather than the mechanism of the transformation it catalyzes, is the critical information on which the design is based. There have already been some examples in which this edge of the tetrahedron has been demonstrated. The de novo design of inhibitors of phospholipase A_2 (Ripka et al. 1987) is perhaps the most striking example to date. A related issue is the design of three-dimensional mimics of naturally occurring ligands, in which the critical elements of the outward structure are incorporated on a simplified framework. The story of the CCK-antagonists reported by the Merck group may be taken as an example of this approach (Evans et al. 1986).

Tendamistat: A Model Target

As a target for mimicry, the 74-amino acid inhibitor of α-amylase known as tendamistat is particularly attractive. Its structure has been determined by X-ray crystallography (Pflugrath et al. 1986) and by 2 D NMR (Kline et al. 1986) and there is excellent agreement between the two methods, particularly with respect to the conformation of the main chain. Moreover, there is evidence that a particular triad of amino acids (Trp_{18}-Arg_{19}-Tyr_{20}) constitutes the "active-loop" of the inhibitor and is important for binding to the target enzyme. We have thus sought to design low molecular weight mimics of this active loop, to demonstrate whether such a strategy for inhibitor design is feasible. The most secure element of the inhibitor structure is the configuration of the backbone and the orientation of the C_α-C_β bonds; these are similar in the crystalline state and in solution and are likely to remain so in the enzyme-inhibitor complex. The relative rigidity of this part of the protein is in contrast to the flexibility of the side chains themselves, which are already seen to be ill-defined in the solution structure and could be significantly reoriented in the enzyme binding site.

We reasoned that the most promising approach would be to design a template which would hold the C_α-C_β bonds of the Trp, Arg, and Tyr sidechains in the appropriate orientation, rather than to design a molecule which would fix the indole, guanidinium, and phenol functional groups themselves in a particular position. The three-dimensional relationship we sought was one between vectors, as opposed to one between Cartesian coordinates. The design of such templates proved to be a challenging task because the number of possibilities to explore is enormous and because the structure input, minimization, and comparison routines require direct interaction with the chemist.

CAVEAT: A Suggestive Program for the Chemist

To assist in the process we have undertaken the development of a computer program, which we call CAVEAT.[3] This program allows the chemist to search a properly constructed structural database for molecules that have substituents whose vector relationship matches that identified in the structure to be mimicked. By suggesting molecules which fit this criterion, the program can provide the chemist with leads in designing a template. The database taken for initial development of the program is the Cambridge Structural Database (CSD), which we have processed from a form that contains essentially Cartesian data into one which contains the necessary vector information. Although still in its initial stage of development, the program has already produced

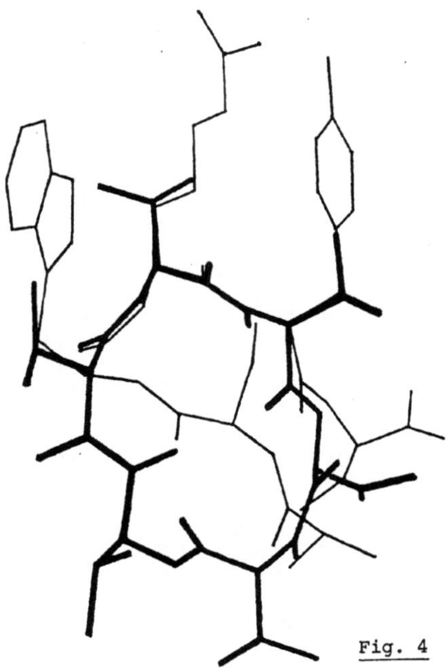

Fig. 4

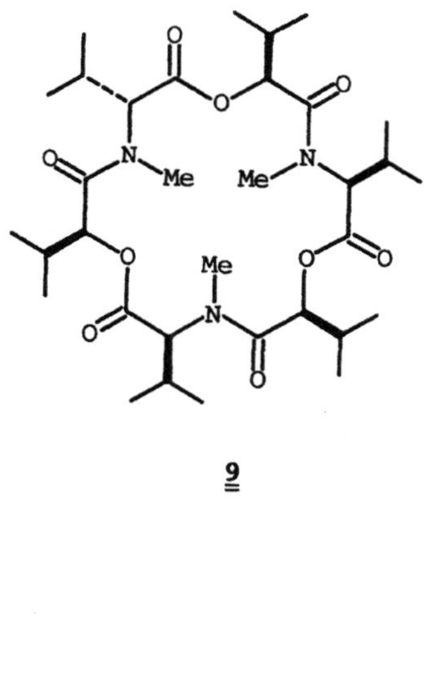

9

Fig. 4. Perspective view of the overlap of cyclodepsipeptide 9 *(heavy lines)* with the C_α-C_β bonds of residues 18-20 from the crystal structure of tendamistat *(thin lines)*

[3]The name of this program is, on the one hand, an acronym derived from "Computer Assisted Vector Evaluation And Target design", and on the other, a reflection of our belief that users of all such programs should maintain a healthy skepticism.

some surprises. For example, a search of the CSD with CAVEAT, using the C_α-C_β bonds of the Trp_{18}-Arg_{19}-Tyr_{20} triad of tendamistat as the criterion, identified the cyclic hexadepsipeptide **9** (Shishova and Simonov 1977), among other possibilities. As revealed in Fig.4, the match between this structure and the triad of tendamistat is remarkable, not only with respect to the C_α-C_β bonds themselves, but even extending to the backbone and the relative orientations of the side chains.

A program that enables the chemist to identify similarities such as this may have a significant impact in the design of mimics, as illustrated for tendamistat, as well as in the invention of molecules to fill vacant binding sites.

Acknowledgments. I would like to express my appreciation to the dedicated students and postdoctoral fellows who are responsible for the work presented above: Charles Marlowe, John Hanson, and Alan Kaplan (zinc peptidase inhibitors), Charles Johnson, Yumi Nakagawa, and Siegfried Reich (synthesis of chorismate mutase inhibitors), and Gary Shea and Stephen Telfer (development of CAVEAT). It has also been a pleasure to collaborate with Professor Brian W. Matthews (University of Oregon) and Professor Peter G. Schultz (Berkeley) and their coworkers on the thermolysin and antibody projects, respectively. Support for much of this work was provided by grants from the National Institutes of Health.

References

Abeles RH (1980) Suicide enzyme inactivators and others. Pure Appl Chem 53:149-160

Abramowitz-Kurn N, Schechter I (1974) Mapping the active site of carboxypeptidase A: a proposed scheme for substrate binding at the active site. Isr J Chem 12:543-555

Bartlett PA, Marlowe CK (1983) Phosphonamidates as transition state analog inhibitors of thermolysin. Biochemistry 22:4618-4624

Bartlett PA, Marlowe CK (1987a) Evaluation of intrinsic binding energy from a hydrogen bonding group in an enzyme inhibitor. Science 235:569-571

Bartlett PA, Marlowe CK (1987b) Possible role for water dissociation in the slow binding of phosphorus-containing transition state analog inhibitors of thermolysin. Biochemistry 27:8553-8561

Bartlett PA, Nakagawa Y, Johnson CR, Reich SH, Luis A (1988) Chorismate mutase inhibitors: synthesis and evaluation of some potential transition state analogs. J Org Chem 53:3195-3210

Christianson DW, Lipscomb WN (1987) Carboxypeptidase A: hydration of a ketonic substrate analogue. Proc Natl Acad Sci USA 84:1512-1515

Evans BE, Bock MG, Rittle KE, DiPardo RM, Whitter WL, Veber DV, Anderson PS, Freidinger RM (1986) Design of potent, orally effective, non-peptidal antagonists of the peptide hormone cholecystokinin. Proc Natl Acad Sci USA 83:4918-4922

Guilford WJ, Copley SD, Knowles JR (1987) On the mechanism of the chorismate mutase reaction. J Am Chem Soc 109:5013-5019

Hangauer DG, Monzingo AF, Matthews BW (1984) An interactive computer graphics study of thermolysin-catalyzed peptide cleavage and inhibition by N-carboxymethyl dipeptides. Biochemistry 23:5730-5741

Holden HM, Tronrud DR, Monzingo AF, Weaver LH, Matthews BW (1987) Slow- and fast-binding inhibitors of thermolysin display different modes of binding: crystallographic analysis of extended phosphonamidate transition state analogues. Biochemistry 26:8542-8553

Jacobs JW, Schultz PG, Sugasawara R, Powell M (1987) Catalytic antibodies. J Am Chem Soc 109:2174-2176

Kline AD, Braun W, Wüthrich K (1986) Studies by ^1H nuclear magnetic resonance and distance geometry of the solution conformation of the α-amylase inhibitor tendamistat. J Mol Biol 189:377-382

Napper AD, Benkovic SJ, Tramontano A, Lerner R (1987) A stereospecific cyclization catalyzed by an antibody. Science 237:1042-1043

Pflugrath JW, Wiegand G, Huber R, Vêrtesy L (1986) Crystal structure determination, refinement and the molecular model of the α-amylase inhibitor Hoe-467A. J Mol Biol 189:383-386

Pollack SJ, Schultz PG (1987) Antibody catalysis by transition state stabilization. Cold Spring Harbor Symp Quant Biol 52 (in press)

Pollack SJ, Jacobs JW, Schultz PG (1986) Selective chemical catalysis by an antibody. Science 234:1570-1573

Rees DC, Lipscomb WN (1982) Refined crystal structure of the potato inhibitor complex of carboxypeptidase A at 2.5 Å resolution. J Mol Biol 169:475-498

Ripka WC, Sipio WJ, Blaney JM (1987) Molecular modeling and drug design: strategies in the design and synthesis of phospholipase A_2 inhibitors. Lect Heterocycl Chem IX:95-104

Sander ME, Witzel H (1985) Direct chemical evidence for the mixed anhydride intermediate of carboxypeptidase A in ester and peptide hydrolysis. Biochem Biophys Res Commun 132:681-687

Schultz PG (1988) The interplay between chemistry and biology in the design of enzymatic catalysts. Science (in press)

Schultz PG, Jackson DY, Jacobs JW, Sugasawara R, Reich SH, Bartlett PA (1988) An antibody-catalyzed Claisen rearrangement. J Am Chem Soc 110:4841-4842

Shishova TG, Simonov VI (1977) Depsipeptide is LLLLLD isomer of Enniatin B. Kristallografiya 22:515-525

Sogo SD, Widlanski TS, Hoara JH, Grimshaw CE, Berchtold GA, Knowles JR (1984) Stereochemistry of the rearrangement of chorismate to prephenate: chorismate mutase involves a chair transition state. J Am Chem Soc 106:2701-2703

Stark GR, Bartlett PA (1984) Design and use of potent, specific enzyme inhibitors. Pharmacol Ther 23:45-78

Tramontano A, Janda K, Lerner R (1986) Catalytic antibodies. Science 234:1566-1570

Tronrud DE, Holden HM, Matthews BW (1987) Structures of two thermolysin inhibitor complexes that differ by a single hydrogen bond. Science 235:571-574

Walsh C (1982) Suicide substrates: mechanism-based enzyme inactivators. Tetrahedron 38:871-909

Wolfenden R (1976) Transition state analog inhibitors and enzyme catalysis. Annu Rev Biophys Bioeng 5:271-306

Evolution of Hemoglobin as Studied by Protein Engineering and X-Ray Crystallography

D. Altschuh,[1,2] J. Tame[1] and K. Nagai[1]

Introduction

Proteins in the globin family are found in a wide variety of organisms, their physiological function being the storage and transport of oxygen. In vertebrate blood, Hb forms a tetramer consisting of two α and two β chains. Assembly into tetramers was a crucial step in globin evolution, allowing it to acquire its allosteric properties. The tetrameric Hb molecule can form two alternative quaternary structures, the T structure with low oxygen affinity and the R structure with high oxygen affinity (Perutz 1979; Monod et al. 1965). The equilibrium between these two structures allows the four subunits to bind oxygen co-operatively and hence Hb can load and unload oxygen efficiently over a small change in partial oxygen pressure. Various metabolites such as H^+, 2,3-diphosphoglycerate (DPG), HCO_3^-, CO_2, and ATP, bind to the T structure more strongly and thereby reduce the oxygen affinity (Kilmartin and Rossi-Bernardi 1973). These effectors are called the heterotropic effectors, and help to deliver oxygen preferentially to actively respiring tissues. Many amino acid residues on the subunit surface are involved in intersubunit contacts and the binding of heterotropic ligands; any mutations at these positions are likely to affect the free energy of the R and T states. This imposes additional selective pressure on the Hb molecule.

Despite this high selective pressure, the globin sequence has diverged considerably and the homology in amino acid sequence between the most distantly related vertebrate Hbs is only 40%. How can such a variety of amino acid sequences give rise to the same three-dimensional fold and cooperative oxygen binding? Different heterotropic allosteric effects have, however, evolved. For example, the oxygen affinity of trout Hb I, one of the Hb components of trout blood, is not affected by H^+ and DPG (Brunori 1975); crocodile Hb has acquired the ability to bind HCO_3^- instead of DPG (Bauer et al. 1981); the alkaline Bohr effect is greatly enhanced in certain fish hemoglobins and is normally referred to as the Root effect (Brunori 1975; Tan et al. 1973). These altered heterotropic effects must have enabled these species to adapt to new environments. If most mutations are functionally neutral, we should be able to introduce these heterotropic effects by 'transplanting' a few residues from other Hbs.

Hemoglobin is an appropriate model to investigate the role of individual amino acid changes in maintaining the structure or altering functional properties; a large number of sequences, as well as high-resolution crystallographic structures, are available so that sequence differences can be analyzed and related to function. We have succeeded in expressing large amounts of human α and β globins in *Escherichia coli* (Nagai et al.

[1]MRC Laboratory of Molecular Biology, Hills Road, Cambridge, England

[2]Institut de Biologie Moléculaire et Cellulaire du CNRS, 15 rue Descartes, 67084 Strasbourg, France

39. Colloquium Mosbach 1988
Protein Structure and Protein Engineering

1985; Tame, Pagnier, Shih, and Nagai, unpublished results). Using our expression system, substitutions can be introduced by site-directed mutagenesis in human hemoglobin. The production of large amounts of the modified protein allows crystallographic studies as well as detailed analysis of function and stability (Nagai et al. 1987; Luisi and Nagai 1986). The interpretation of results should be somewhat easier for a carrier protein like hemoglobin than for an enzyme.

Effects of 'Conservative' Amino Acid Replacements

The amino acid sequences of the majority of proteins have undergone considerable change during the evolution of species (Dayhoff 1972). It is widely believed that most of these amino acid replacements are simply the result of the chance fixation of functionally neutral or nearly neutral mutations (Kimura 1979) and that Darwinian selection played an important role only in excluding the deleterious mutations. This theory of molecular evolution is based solely on the comparison of protein sequences from different species. There is clearly an enormous difference in the interchangeability of amino acids in a protein molecule depending on their chemical and structural environment. Some positions are absolutely conserved; some have accepted only coservative mutations (replacements with chemically and structurally similar amino acids); whereas others, mainly on the molecular surface, have accepted a variety of amino acids. The average frequency of each amino acid replacement has been calculated from a statistical analysis of protein sequences (Dayhoff 1972) but the substitution of individual residues largely depends on their micro-environment. For example, Val at some positions is exclusively replaced by Ile, but at other positions only by Leu. Similarly, some Leu residues are exclusively replaced by Phe and others only by Met. Can differences in local environment account for the interchangeability at each position? It is not known what proportion of mutations are functionally neutral, since there are only limited experimental data on the effect of individual evolutionally accepted mutations on the function, structure, and stability of proteins. We have studied amino acid replacements in the oxygen binding pocket of hemoglobin where the effect of each mutation is directly reflected by the oxygen affinity and rate of autoxidation.

Substitutions at the Oxygen Binding Pocket. Oxygen binds to the heme group which is embedded in the hydrophobic pocket of the globin. His-F8 forms a covalent bond to the heme iron and is absolutely conserved. The oxygen binding site is buried within the densely packed protein interior and there is no room for oxygen in the β-subunit of deoxy Hb. The oxygen binding site of the α-subunit is somewhat larger and a water molecule is hydrogen bonded to His-E7α when no ligand is present (Fermi et al. 1984). In oxyHb, His-E7 and Val-E11 are in van der Waals contact with the oxygen ligand, and so volume changes of these residues can affect the geometry of the Fe-O_2 bond (Shaanan 1983). Of about 100 known Hb sequences, Val-E11β is invariant except in one species, where it is replaced by Ile. In the α-subunit, Ile is found at E11 in about 10% of known sequences. Does the interchangeability of Val-E11 correlate with its micro-environment in each subunit? Why are Leu and Met, hydrophobic residues of similar size to Val and Ile, not found at the E11 position? If we introduce these residues by site-directed mutagenesis, do they have a measurable effect on function or stability? In order to answer these questions, we have replaced Val-E11 of both the α- and β-subunits with Ile, Leu, Ala, and Met and measured various functional parameters. We have determined the structure of these mutants by X-ray crystallography to see how new side chains are accom-

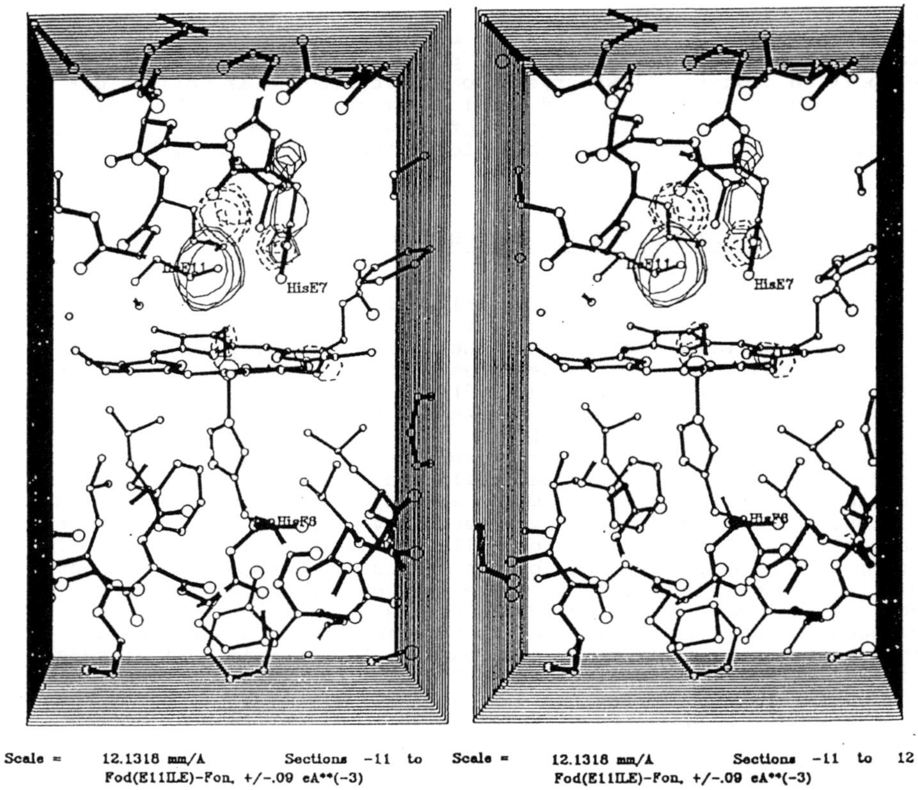

Fig. 1. The structure of the oxygen binding site in the Val-E11β→Ile mutant. A difference Fourier map of the mutant minus Hb A is shown. A large positive peak is due to the group of Ile-E11β which covers the oxygen binding site and hinders oxygen binding. A pair of positive and negative peaks around His-E7β and Leu-B10β shows that these residues have been displaced. Positive contours (——); negative countours (---)

modated and the nature and extent of the structural adjustments required (Nagai et al. 1987).

Crystallographic Analysis of Mutants

The Val-E11β→Leu, Ile Mutations. Figure 1 shows a difference Fourier map of the Ile-E11β mutant minus Hb A at 2.45 Å resolution. In an α-helix, the forked side chain of a Val or Ile residue cannot rotate about the C_α-C_β bond. Therefore the δ-methyl group of the Ile side chain is forced to cover the oxygen binding site and hinder oxygen binding. Pairs of positive and negative peaks next to His-E7β and Leu-B10β indicate that these residues are displaced by the Ile-E11β side chain. This can be done by small adjustments of the side chain torsion angles. The difference Fourier map of the Leu-E11β minus Hb A at 2.18 Å resolution shows a large positive peak due to $C_{\delta 1}$ and $C_{\delta 2}$ of the new Leu side chain and two negative peaks due to loss of $C_{\gamma 1}$ and $C_{\gamma 2}$ of Val-E11β (Fig. 2).

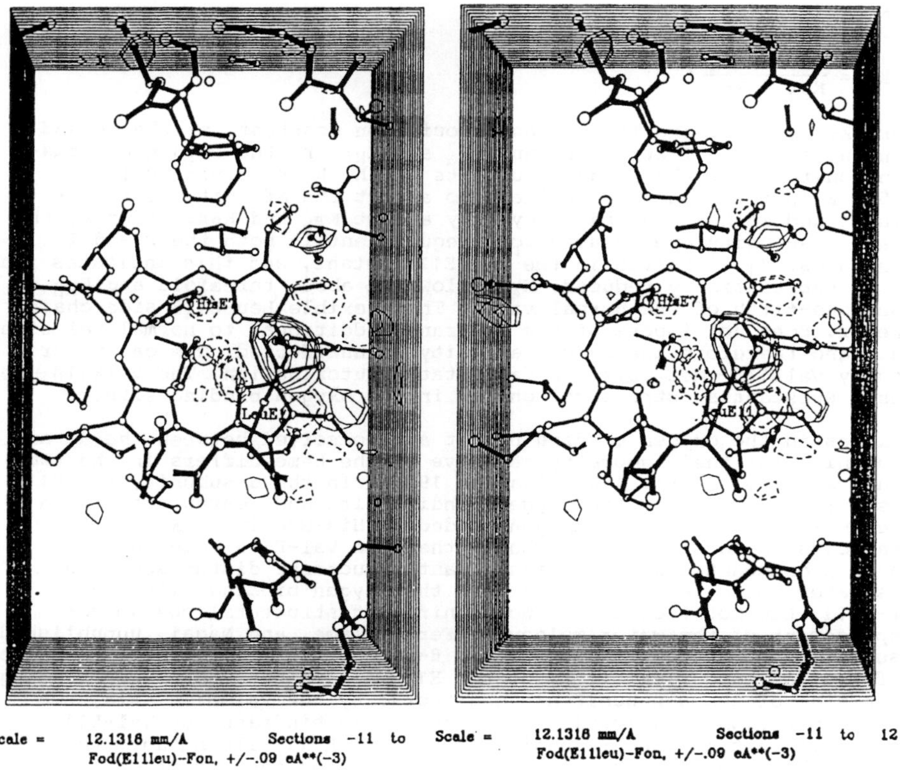

Fig. 2. A difference Fourier map of the Val-E11β → Leu mutant minus Hb A. The Leu side chain is as big as the Ile side chain but it is more flexible. The Leu-E11β exercises no additional steric hindrance over the normal Val side chain and therefore the oxygen affinity of the Leu-E11β mutant is normal. Positive contours (——); negative contours (---). (Nagai et al. 1987)

The Leu-E11β side chain neither extends to the oxygen binding site nor displaces His-E7β, but causes a movement of Leu-B10β away from Leu-E11β. The Leu side chain can rotate about the C_α-C_β and C_β-C_γ bonds is therefore flexible. In fact, the Leu-E11β side chain must be disordered since we could not fit it into the 2F mutant-F-native map with a single conformation. The oxygen equilibrium curve of the Leu-E11β mutant is indistinguishable from that of native Hb A. Owing to its flexibility, the Leu-E11β side chain exercises no additional steric hindrance over the normal Val-E11β side chain. However, the Ile-E11β side chain, which covers the oxygen binding site, lowers the oxygen affinity significantly. In the generalized allosteric model which allows the oxygen affinity of the α- and β-subunits to be different, K_1 (the first Adair constant: binding constant for the oxygen) and K_4 (the fourth Adair constant: binding constant for the fourth oxygen) are given by the following equation:

$$K_1 = (1/2)(K_{T\alpha} + K_{T\beta})$$

$$K_4 = \frac{2K_{R\alpha}K_{R\beta}}{K_{R\alpha} + K_{R\beta}}$$

where $K_{T\alpha}$ and $K_{T\beta}$ are the oxygen association constants of the α- and β-subunits in the T structure and $K_{R\alpha}$ and $K_{R\beta}$ are the oxygen association constants of the α and β subunits in the R structure (Baldwin 1975). E11β mutations are unlikely to affect the affinity of the α-subunit and we assumed that they only affect $K_{T\beta}$ and $K_{R\beta}$. In Hb A, the α- and β-subunits are assumed to be equivalent in both the R and T structures. K_1 is halved in the Ile-E11β mutant, and this indicates that $K_{T\beta}$ is substantially reduced. K_4 is lowered about threefold and $K_{R\beta}$ is about one-fifth of the normal value. The flexible Leu-E11β side chain, however, does not impose steric hindrance additional to normal Val side chain and therefore the oxygen affinity is unaffected. The cavity created by ValE11β→Met mutation facilitates autoxidation. The ValE11β→Met causes substantial structural change in the oxygen binding site.

Residues in the oxygen binding pocket are conserved between the α- and β-subunits but their geometry relative to the heme differs in the two subunits (Fermi et al. 1984; Shaanan 1983). In the α-subunit, Val-E11 lies further away from the oxygen binding site and leaves room for a water molecule which is hydrogen bonded to His-E7α in deoxy Hb. Our X-ray crystallographic study showed that the Val-E11α→Ile mutation can be accommodated without significant structural disturbance, but this water molecule is excluded from the oxygen binding site. The Val-E11α→Leu mutation causes no significant structural change nor displacement of the water molecule (Fermi, Tame, and Nagai, unpublished results). The oxygen affinity of the β-subunit is sensitive to the size, conformation, and flexibility of the E11 residue but the oxygen binding properties of the α-subunit are unaffected by the Val-E11α→Ala, Leu and Ile mutations, indicating that the steric hindrance by Val-E11 is not significant in the α-subunit. In the Val-E11α→Ile mutant, the extra δ-methyl group of the Ile may compensate for the effect of loss of the water molecule (Tame, Pagnier, Shih, and Nagai, unpublished results).

The His-E7β→Gln Mutation. His-E7 is highly conserved in myoglobin and hemoglobin. A neutron diffraction study of the oxy Mb has shown that the oxygen ligand is hydrogen bonded to His-E7 (Phillips and Schoenborn 1981). Such a study has not been carried out for oxy Hb, but the X-ray crystallographic structure at 2.1 Å suggests a hydrogen bond between the oxygen molecule and His-E7 in both subunits although it is somewhat weaker in the β-subunit (Shaanan 1983). His-E7 is invariant except in the α-subunit of opossum Hb where it is replaced by Gln (Stenzel et al. 1979). Figure 3 shows the electron density map of the His-E7β→Gln mutant at 2.18 Å resolution. The new side chain is well-ordered but we cannot distinguish between N_ε and O_ε of the Gln-E7β side chain at this resolution. The side chain conformation is consistent with the presence of a hydrogen bond. The flexible Gln side chain can donate a hydrogen bond to the oxygen molecule with reasonable geometry. In this sense, this replacement is conservative but the oxygen affinity of this mutant is significantly higher than that of Hb A. The replacement of His-E7 with Val, Gly, and Phe facilitates autoxidation and makes Hb very unstable. It is not clear why hydrophobic groups cannot be accepted at this position, but our resonance Raman study showed that the ν(C-O), ν(Fe-CO) and δ(Fe-C-O) vibrational modes of CO Hb are affected when His-E7β is replaced by Gly and Val but not by Gln. This indicates that the polarity of E7β side chain affects the electronic state of the Fe-ligand bond.

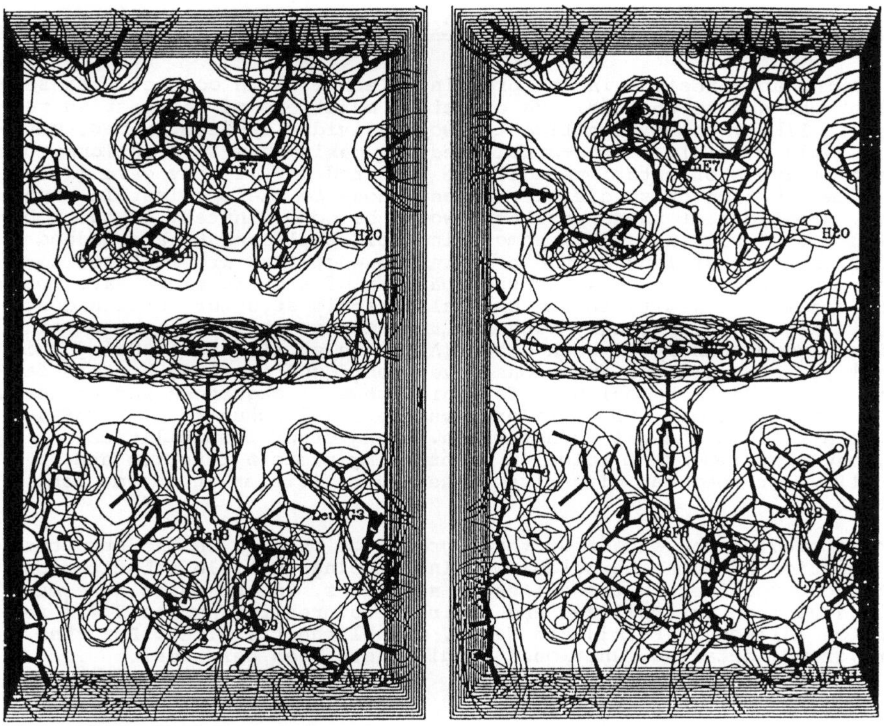

Scale = 12.1318 mm/A Sections -10 to
2Fdiff-Fcalgln, 0.25 eA**(-3)

Scale = 12.1318 mm/A Sections -10 to 6
2Fdiff-Fcalgln, 0.25 eA**(-3)

Fig. 3. The structure of the His-E7β → Gln mutant. It shows that N_ε of Gln-E7 side chain can form a hydrogen bond to oxygen as does the N_ε of His-E7. (Nagai et al. 1987)

Evolutionary Considerations. Our experiment sheds some light on the interchangeability of amino acids during evolution. In the densely packed protein, small differences in the size, flexibility, and conformation of different residues can affect the function and stability of the protein. In both subunits, the replacement of Val-E11 with smaller Ala greatly facilitates autoxidation and makes the molecule unstable. The replacement of Val-E11 with Ile introduces an extra methyl group into the oxygen binding pocket. This causes some structural change in the β-subunit and reduces the oxygen affinity. However, the same mutation in the larger pocket of the α-subunit has no significant structural or functional effect. The extra methyl group simply displaces the water molecule hydrogen-bonded to His-E7α from the heme pocket. This is reflected in the difference in the frequency of the Val-E11 → Ile mutation between the α- and β-subunits during evolution. Only one species accepted this mutation in the β-subunit whereas 10% of known Hb sequence have this mutation in the α-subunit. In both subunits, the Val-E11 → Leu mutation does not affect the oxygen binding properties nor the rate of autoxidation significantly, but Leu has not been found at this position. It is not obvious why this mutation is excluded at this position.

The Co-Ordinated Mutations in Protein Evolution

Each amino acid residue in the protein molecule is in contact with several others and so the additive effect of mutations of interacting residues is likely to be significant. Two, individually deleterious, mutations could possibly be "co-stabilized" by making new interactions, or an entire group of mutations could be selected for if they gave rise to a new function or specificity advantageous to the species. In such cases, mutations at these positions would be co-ordinated. In the tobamovirus coat protein family, some amino acid replacements are indeed co-ordinated: positions in the protein sequence that are replaced together, and therefore show the same pattern of substitutions, are mostly in spatial proximity in the crystallographic structure of the coat protein disk. This strongly suggests that these mutations are not stabilized independently of each other (Altschuh et al. 1987). Using this principle, co-ordinated mutations have been found in three protein families other than tobamoviruses for which several sequences and at least one crystallographic structure are available (Altschuh et al. 1988). The role of such changes is not known. Among other possibilities, they may play important roles in maintaining protein function and stability, or they may give rise to new specificities in enzyme families.

We have studied the effect of three groups of changes on the structure and function of human hemoglobin by introducing multiple substitutions that occur together in the different sequences. In two cases, the mutants have been crystallized and their structures refined to 2.4 Å resolution (Altschuh, Fermi, and Nagai, unpublished results). For the third group, detailed functional studies have been performed.

Method for Detecting Co-Ordinated Changes in Protein Families

Co-ordinated amino acid replacements were searched for in hemoglobin using the method described by Altschuh et al. (1987). This method requires several homologous protein sequences and the crystallographic structure of at least one of these proteins. In the first step, groups of residues with an identical pattern of amino acid changes are constructed from the sequence alignment as shown in Table 1. The spatial location of residues with an identical pattern is then examined to see whether they are close in the crystallographic structure. Optimal conditions for detecting such changes are high structural constraints and low sequence homology (Altschuh et al. 1988). The stringent original method failed to pick up some pairs of residues which are known to interact in human Hb. We have examined how these pairs of residues are mutated in different globin sequences (Altschuh and Nagai, unpublished results). The following modifications are required to analyze the pattern of amino acid substitutions in closely related proteins such as hemoglobin.

1. Some amino acid types are considered alternatively non-equivalent or equivalent (for example Phe-Leu, Lys-Arg).

2. Positions with a similar but not identical substitution pattern are selected: some exceptions are allowed.

3. In regions with low structural constraints, residues that interact in a given structure may form different contacts after substitution. Such positions are expected to be conserved together in some sequences and mutate independently in other sequences. To detect such residues,

Table 1. Schematic explanation of the method used for detecting co-ordinated amino acid changes in homologous proteins

1. Alignment of four sequences containing eight positions

Residue position		*1*	*2*	*3*	*4*	*5*	*6*	*7*	*8*
Reside type in sequence	*1*	K	S	L	E	C	T	F	S
	2	K	N	L	D	G	T	F	S
	3	H	N	V	T	G	V	M	A
	4	H	S	F	Q	S	V	M	T

2. The substitution pattern (read vertically) at each position is symbolized by four numbers

Residue position		*1*	*2*	*3*	*4*	*5*	*6*	*7*	*8*
Sequence	*1*	1	1	1	1	1	1	1	1
	2	1	2	1	2	2	1	1	1
	3	2	2	2	3	2	2	2	2
	4	2	1	3	4	3	2	2	3

3. Search for positions having an identical pattern of amino acid substitutions

Patterns represented by more than one sequence position are:

Pattern 1 1 2 2	Positions 1,6,7
Pattern 1 1 2 3	Positions 3,8

4. Analysis of spatial proximity between two or more positions from each group using the crystallographic structures available

the sequence alignment is searched for positions that contain the same amino-acid type in a given sub-set of sequences, regardless of what happens in the remaining sequences. Residues that are conserved in the same set of sequences include, for example pairs, that are known to form a salt bridge in human hemoglobin. When amino acids which form salt-bridges are replaced by uncharged residues, their partners are no longer selectively maintained and are also replaced by uncharged residues (Altschuh and Nagai, unpublished results).

An alignment of 56 hemoglobin sequences has been analyzed as described above. Three interesting examples of co-ordinated mutations both in the internal residues and surface residues are presented.

1. The amino acid substitution pattern of positions 74β and 133β shows the best correlation (one exception out of 56 sequences analyzed). The pair Gly-74β+Val-133β occurs in mammals only; the pair Ala-74β+Leu,Phe-133β in all other sequences. Gly-74β is not in contact with Phe-133β but is part of the E-helix, whose side chains, in particular Phe-71β, Leu-75β and Leu-78β, interact with Phe-133β. These substitutions represent an increase in side chain volume.

2. The second group of changes is Ser-102α → Gly and Leu-105α → Phe. The substitution pattern for these two positions is correlated in 50 out of 56 sequences analyzed. Mammal (except guinea pig) and kangaroo sequences have Ser-102α and Leu-105α; bird sequences have Gly-102β and Phe-105β (except 102 Ala in flamingo); other sequences have mostly Ala-102β and Val,Ile-105β. Positions 102β and 105β are buried and located close to each other on two successive helix turns (minimal distance between side chain atoms is 5.7 Å).

3. The pattern of substitution of surface residues is often complicated but changes at positions 90β, 94β and 144β are very closely correlated. These positions are conserved in mammalian and bird sequences but con-

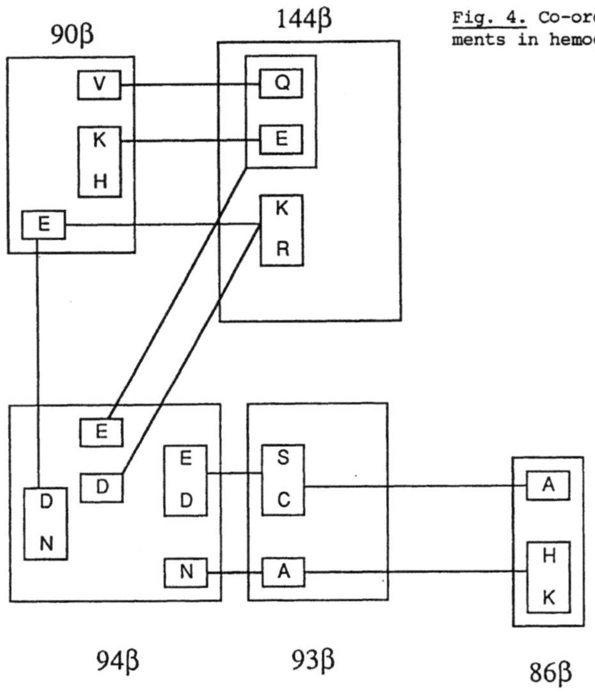

Fig. 4. Co-ordinated amino acid replacements in hemoglobin

tain different amino acid types in reptile, amphibian and fish sequences. Some amino acid types at neighbouring positions 87β and 93β can also be related to the substitution pattern at these positions. Figure 4 summarizes the relation between amino acid types at positions 87β, 90β, 93β, 94β, and 144β.

Some of these positions are believed to play an important part in the Bohr effect. In order to investigate their roles, we have introduced the residues found in fish hemoglobin into human one at a time by site directed mutagenesis.

The Bohr Effect. Within physiological conditions the oxygen affinity of Hb decreases as the pH is lowered, a phenomenon called the alkaline Bohr effect. Deoxy Hb is more positively charged than oxy Hb since protons are released upon oxygenation. Protons, in turn, stabilize deoxy Hb relative to oxy Hb and hence reduce the oxygen affinity. The Bohr effect can be quantified as the number of protons released upon oxygenation or as a change in oxygen affinity per unit pH change. Asp-94β forms a salt bridge with His-146β in deoxy Hb so that the pK value of this imidazole group is increased to pH 8.0 (Perutz 1979; Kilmartin and Rossi-Bernardi 1973). This salt bridge is broken upon oxygenation and the imidazole group of His-146β returns to its normal pK value and hence releases Bohr protons. His-146β is responsible for about 40 - 50% of the alkaline Bohr effect in human Hb. Asp-94β is strongly conserved but is replaced by Glu in some reptile, amphibian and fish Hb. As shown in Fig.4, whenever 94β is Glu, Glu-90β is always replaced either by a positively charged or an uncharged residue. The residues at 90β and 144β show very closely co-ordinated mutations. Most species have Glu at 90β and Lys or Arg at 144β. Nile crocodile, alligator, and caiman

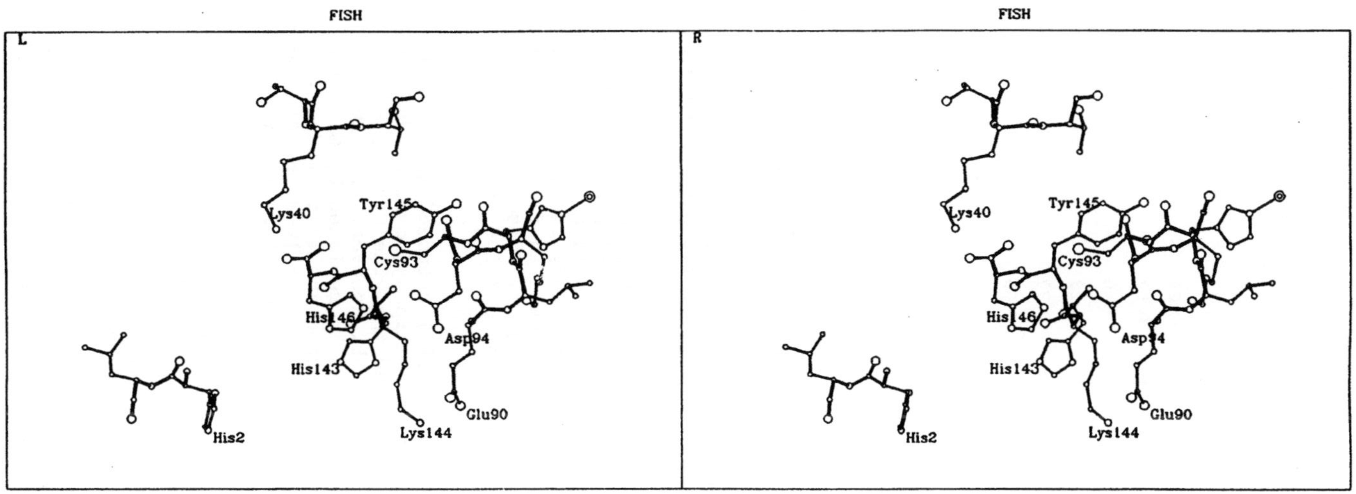

Fig. 5. Structure of deoxy Hb based on Fermi et al. (1984)

Table 2. Effect of step-by-step mutations on the Bohr effect and the DPG effect[a]

Mutant	2β	90β	93β	94β	143β	144β	Bohr effect	DPG effect
Human	His	Glu	Cys	Asp	His	Lys	0.50	0.44
A	His	Glu	Ser	Asp	His	Lys	0.47	0.45
B	His	Glu	Ser	Glu	His	Lys	0.24	ND
C	His	Glu	Ser	Glu	His	Gln	0.27	0.55
D	His	Val	Ser	Glu	His	Gln	0.46	0.55
E	Glu	Val	Ser	Glu	His	Gln	0.23	0.24
F	Glu	Val	Ser	Glu	Arg	Gln	0.77	0.56
Fish	Glu	Val	Ser	Glu	Arg	Gln	0.85	0.50

[a]ND, not determined.
Human Hb A releases 0.5 protons per heme upon oxygenation. His-146β is responsible for 40% of the alkaline Bohr effect. The Bohr effect is measured as $\Delta \log P_{50}/\Delta pH$; the DPG effect as $\log P_{50}$ (with 2 mM DPG)-$\log P_{50}$ (with no DPG) (Tame, Altschuh, Poyart, and Nagai, unpublished results).

Hbs have Lys at 90β and Glu at 144β. South American lungfish Hb has His at 90β and Glu at 144β. The alkaline Bohr effect is almost twice as large in some fish Hbs as in human Hb and is called the Root effect. At low pH the molecule remains in the T structure even when fully oxygenated and thus oxygen binding becomes non-cooperative. Carp, goldfish, and trout Hb IV, which all show the Root effect have uncharged amino acids at both positions 90β and 144β.

Figure 5 shows an atomic model of deoxy Hb which includes residues 90β, 93β, 94β, 143β, and 144β. Glu-90β, Lys-144β and Asp-94β are spatially close but no direct interactions, such as a salt-bridge or a hydrogen bond have been found between them. In order to study the role of these amino acids, we have introduced the residues found in fish Hb into human Hb one at a time by site-directed mutagenesis. These mutations are the Cys-93β → Ser, Asp-94β → Glu, Lys-144β → Gln, Glu-90β → Val, His2β → Glu, His-143β → Arg mutations.

Step-by-Step-Mutagenesis. Perutz and Brunori (1982) proposed that Ser-93β could stabilize the T structure by forming extra hydrogen bonds with the amide and carbonyl group of His-146β and that the Cys-93β → Ser would introduce the Root effect into Human Hb. However, this mutation had only a small effect. In fish Hb, Asp-94β is replaced by Glu. The longer Glu side chain was thought to strengthen the salt-bridge with His-146β but the additional Asp-94β → Glu mutation halved the Bohr effect (Table 2). Although the Asp-94β → Glu replacement is seemingly conservative, a salt-bridge is not formed between His-146β and Glu-94β. Our X-ray crystallographic analysis of this mutant showed that these two residues are disordered. The salt bridge is present in fish Hb and therefore additional mutations are required to favor its formation. We then introduced the Lys-144β → Gln and Glu-90β → Val mutations since the Asp-94β → Glu mutation is co-ordinated with these two substitutions. The alkaline Bohr effect is indeed restored by these additional mutations. This clearly indicates that these three mutations are not independent. The neutrality of the replacement at 94β depends on the surrounding residues. We do not know the order in which these mutations arose during evolution but tuna fish Hb, which retains the Root effect, has Val at 90β, Asp at 94β and Lys at 144β and is probably an intermediate. The tuna fish sequence indicates that none of these replacements is essential for the Root effect. All the fish Hb which show the Root effect have Glu at 2β and Arg at 143β and we also introduced these mutations. The His-2β → Glu mutation reduced the alkaline Bohr effect by 50% but the additional His-143β → Arg mutation increased it to 170% of the normal Hb A value. His-2β is involved in the DPG binding and the His-2β → Glu mutation halved the DPG effect but this is compensated by changing His-143β → Arg. Although we could engineer an enhanced Bohr effect into human Hb, the oxygen affinity of our final mutant is much higher than that of fish Hb. Many additional mutations are probably required to reduce the overall oxygen affinity.

Co-Ordinated Mutations and Neutral Mutations. The neutral theory of evolution proposed by Kimura (1979) states that the majority of amino acid changes which protein sequences have undergone during species evolution are functionally neutral or nearly so. We have tested this hypothesis by trying to engineer new or altered functions in human Hb. If the neutral theory is correct, then it should be possible to confer these changes by altering only a small number of key residues. The function we have tried to "evolve" in human Hb is the Root effect, a marked pH dependence of ligand affinity. As shown above, the Asp-94β → Glu mutation halves the Bohr effect of human Hb and the additional mutations Glu-90β → Val and Lys-144β → Gln restore it. Thus the seemingly conservative Asp → Glu mutation is only functionally neutral if neighbouring residues are also changed. Similarly, the His-2β → Glu mutation reduces the ability of DPG to alter the quaternary state equilibrium, but the additional His-143β → Arg mutation has a compensatory effect.

Our experiments suggest that a protein function may be maintained by a wide variety of co-ordinated mutations which are not individually functionally neutral. Co-ordinated mutations, which are accompanied by changes in residue volumes, are found in the densely packed protein interior (see example 1). Both structural and functional studies are underway to investigate the evolutionary significance of these mutations (Altschuh, Fermi, and Nagai, unpublished results).

Acknowledgments. The authors thank Drs. Giulio Fermi, Ben Luisi, Daniel Shih, Claude Poyart for their contributions and Dr. Max Perutz for his encouragements.

References

Altschuh D, Les KAM, Bloomer AC, Klug A (1987) Correlation of co-ordinated amino acid substitutions with function in viruses related to tobacco mosaic virus. J Mol Biol 193:693-707

Altschuh D, Vernet T, Berti P, Moras D, Nagai K (1988) Coordinated amino acid changes in homologous protein families. Protein Engineering 2:193-199

Baldwin JM (1975) Structure and function of haemoglobin. Prog Biophys Mol Biol 29: 225-320

Bauer C, Foster M, Gros G, Mosca A, Perella M, Rollema HS, Vogel D (1981) Analysis of bicarbonate ion binding to crocodilian hemoglobin. J Biol Chem 256:8429-8435

Brunori M (1975) Molecular adaptation to physiological requirement: hemoglobin system of trout. Curr Top Cell Regul 9:1-39

Dayhoff MO (1972) Atlas of protein sequence and structure. National Biomedical Research Foundation, Washington, D.C.

Fermi G, Perutz MF, Shaanan B, Fourme R (1984) The crystal structure of human deoxyhaemoglobin at 1.74 Å resolution. J Mol Biol 175:159-174

Kilmartin JV, Rossi-Bernardi L (1973) Interaction of hemoglobin with hydrogen ions, carbon dioxide and organic phosphate. Physiol Rev 53:836-890

Kimura M (1979) The neutral theory of molecular evolution. Sci Am 241:94-104

Luisi B, Nagai K (1986) Crystallographic analysis of mutant human haemoglobin made in *Escherichia coli*. Nature 320:555-556

Monod J, Wyman J, Changeux J-P (1965) On the nature of allosteric transitions: A plausible model. J Mol Biol 12:88-125

Nagai K, Perutz MF, Poyart C (1985) Oxygen binding properties of human mutant hemoglobins synthesized in *Escherichia coli*. Proc Natl Acad Sci USA 82:7252-7255

Nagai K, Luisi B, Shih M, Miyazaki G, Imai K, Poyart C, De Young A, Kwiatokowski L, Noble RW, Lin S-H, Yu N-T (1987) Distal residues in the oxygen-binding site of haemoglobin studied by protein engineering. Nature 329:858-860

Perutz MF (1979) Regulation of oxygen affinity of hemoglobin. Influence of structure of the globin on the heme iron. Annu Rev Biochem 48:327-386

Perutz MF, Brunori M (1982) Stereochemistry of cooperative effects in fish and amphibian haemoglobins. Nature 299:421-426

Phillips SEV, Schoenborn BP (1981) Neutron diffraction reveals oxygen-histidine hydrogen bond in oxymyoglobin. Nature 292:81-82

Shaanan B (1983) Structure of human oxyhaemoglobin at 2.1 Å resolution. J Mol Biol 171:31-59

Stenzel P, Brimhall B, Jones RT, Black JA, McLachlan A, Gibson D (1979) Opossum hemoglobin: The amino acid sequences of the α- and β-chains. J Biol Chem 254: 2071-2076

Tan AL, Noble RW, Gibson QH (1973) Conditions restricting allosteric transitions in carp hemoglobin. J Biol Chem 248:2880-2888

Design and Construction of Biologically Active Peptides and Proteins Including Enzymes

E. T. Kaiser[1]

Introduction

At the present time, peptides up to 30 or 40 amino acids in length are readily accessible through the Merrifield solid-phase method (Barany and Merrifield 1980). The major strength of this method is that it is not necessary to purify the intermediate peptides in the course of synthesis and that very high yields can be obtained in the various coupling steps. With the great improvements that have occurred in high pressure liquid chromatography (HPLC) the purification of the target 30 or 40 amino acid-long peptide from impurities such as deletion analogs has become quite feasible. When one considers the preparation of larger peptides (50 amino acids or more) or small proteins (in the vicinity of 100 amino acids), it becomes far less practical to carry out preparations by the stepwise solid-phase method. In particular (Clark-Lewis et al. 1986), the separation of incomplete peptides from the target peptide becomes much more difficult as the length of the peptide grows and, while there have been reports of the synthesis of materials greater than 100 amino acids in length, the purification of these becomes very difficult, and it becomes extremely hard to make a case for the purity of the final product. Of course, one could argue that for peptides which are in the range of 50 to 100 amino acids in length the best solution for their preparation would be to use molecular biological methods. That is, in principle, one should be able either to clone or to synthesize the gene for the appropriate peptide or small protein and then express the target peptide in a suitable host such as *E. coli* or yeast. This is not, however, always a satisfactory solution. Sometimes, expression of the target peptide or protein is not straightforward for reasons which could include proteolysis by enzymes in the host medium. Furthermore, total synthesis of peptides or proteins has several advantages relative to the use of molecular biological methods. By the synthetic approach one can introduce D-amino acids, unnatural amino acids and even nonpeptidic regions in the peptide produced. By the molecular biological approach it would be very difficult to introduce a specific label such as ^{13}C or ^{15}N at a particular location without labeling all other amino acids of a given type. Of course, by chemical methods it should be quite feasible to introduce a single label wherever one wants it.

How Can Peptides Which Are from Fifty to Approximately One Hundred Amino Acids in Length Be Synthesized?

Even if it is granted that there is a need to develop chemical methods for the preparation of peptides 50 to 100 amino acids in length, the question must be asked as to how this synthesis might be accomplished.

[1]The Rockefeller University, 1230 York Avenue, New York, NY 10021-6399, USA

39. Colloquium Mosbach 1988
Protein Structure and Protein Engineering

In recent times we have been exploring what appears to be a rather powerful approach to this problem. Several years ago, we developed a new resin, a polystyrene-1% divinylbenzene copolymer which we derivatized with p-nitrobenzoyl chloride. The resultant p-nitrobenzoyl-resin was treated with hydroxylamine to give the corresponding polymer bound oxime (DeGrado and Kaiser 1980,1982). The first amino acid of the growing peptide chain was introduced by coupling an N^{α}-protected amino acid with the oxime resin using dicyclohexycarbodiimide as the coupling reagent. Subsequently, the N^{α}-protecting group was removed from the amino terminus of the resin bound amino acid and a cycle of coupling of N^{α}-protected amino acids to the material bound on the polymer, followed by deprotection using trifluoroacetic acid was carried out. The protected peptide segment which was our objective was then obtained by treating the resin-bound peptide with N-hydroxypiperidine and the ester obtained was treated further with Zn and acetic acid to give the C-terminal carboxyl group in a free form. The latter treatment to give the free acid is feasible for any peptide which does not contain a sulfur-containing function such as Met.

The protected peptide segments isolated from the oxime approach can be purified readily on a variety of columns and, generally, we carry out a final purification by HPLC. The segments can be characterized by typical chemical methods including mass spectroscopy and amino acid analysis.

Synthesis of Apolipoprotein Models

We have assembled these sorts of small protected segments either by sequential coupling of them on a solid support such as the oxime resin or by solution coupling. Among the peptides and proteins which we have synthesized by this approach are 22- and 44-amino acid peptide models of apolipoprotein A-I (apo A-I) (Nagakawa and Kaiser 1983,1985), the principal protein constituent (243 amino acids in length) of high density lipoprotein. The 22-peptide has a sequence which allows it to assume an "idealized" structure mimicking the prototpyic amphiphilic α-helical regions of apo A-I. It is thought that in the apo A-I molecule there are roughly six to seven such amphiphilic α-helical regions (Fitch 1977) and it has been suggested that these regions lie on the surface of the HDL particle with the axes of the helices lying approximately tangential to the surface and the hydrophobic face of the helix penetrating somewhat into the particle between the phospholipid head groups (Kaiser and Kezdy 1984). Our model 22-peptide was constructed principally out of Leu residues as the hydrophobics, Lys as the positively charged hydrophilics and Glu as the negatively charged hydrophilics. These residues are in their respective categories the best α-helix formers. In the model peptide, the hydrophobic-hydrophilic balance correpsonds closely to what we have distilled as a prototypic structure from the examination of the putative α-helical regions of apo A-I and the charge balance is also similar (Fukushima et al. 1979). However, the amino acid sequence homology between the model and any of the α-helical segments of apo A-I has been reduced as much as possible. Nevertheless, the model 22-peptide simulates very well the physical and biological features of the apo A-I molecule, despite the fact that it is only 1/11th the length of the whole protein.

In the apo A-I molecule, it is notable that the putative amphiphilic α-helical segments are connected through "hinges" containing either Pro or Gly residues. From our model building, it appeared to us that such hinges could play an important role in adjusting the structure of apo A-I to the curved surface of the HDL particle. We undertook to synthesize a covalent dimer of the 22-peptide model of apo A-I. This was readily accomplished by the combination of segment synthesis and condensation, and we were able to use the principal intermediates in the preparation of the 22-peptide for the synthesis of the 44-peptides (Nakagawa and Kaiser 1985). We found that the 44-peptide was an even better model for apo A-I than had been the 22-peptide. For the 44-peptide, when the limiting area per residue was measured either in a Langmuir trough at the air-water interface or in studies of binding to polystyrene beads coated with phospholipid, these areas were found to be identical to that measured for apo A-I itself. Also, the characteristics of the 44-peptide in terms of its α-helicity measured in different solvent systems by circular dichroism spectroscopy was very similar to the behavior of the apo A-I molecule.

Preparation of Larger Peptides by Segment Synthesis-Condensation

Having demonstrated through our work with the 22- and 44-model peptides that our segment synthesis-condensation approach is very viable, we turned to larger peptides which would be more difficultly accessible by stepwise solid-phase synthesis. A particularly exciting example of a molecule which we have succeeded in preparing (Mihara and Kaiser 1988) is a 60 amino acid peptide corresponding to the homeo domain region of *Antennapedia* protein, a protein involved in *Drosophila* development. This peptide has been prepared in good yield and in high purity, as judged by a variety of techniques including automated Edman degradation. Footprinting experiments show that the homeo domain binds to the AAT-rich DNA sequence in the Antp gene. Many of the physical and biological properties of our peptide are currently under investigation.

Another important system which we have snythesized recently is the RNase T_1 molecule, as well as a structural analog. While the material we have obtained in our segment synthesis-condensation approach has not been brought to full purity, we have already shown that our RNase T_1 sample is catalytically active in degradation of yeast RNA (Sasaki and Findeis, personal communication). Additionally, we have replaced an amphiphilic α-helix present in the RNAse T_1 molecule by a designed helix employing an approach similar to what we used for the apo A-I models. Again, we have not purified the mutant RNAse T_1 molecule to homogeneity, but it is clear that it is quite catalytically active, and it appears, therefore, that we have been able to redesign a significant structural unit in a folded protein and still obtain what appears to be proper folding. Since the redesign involved changing 11 out of the 18 amino acids in the helical region, we feel that we have accomplished an important first step towards the design of tertiary structure from primary amino acid sequence.

Acknowledgments. Deep appreciation is expressed for the effort of the many coworkers whose names are cited in the references I have given. The partial support of this research by NIH Program Project HL18577 is gratefully acknowledged.

References

Barany G, Merrifield RB (1980) In: Peptides, analysis, synthesis, biology, vol 2, Gross E, Meienhofer H (eds) Academic Press, NY, pp 1-284
Clark-Lewis I, Aebersold R, Ziltener H, Schrader JW, Hood LE, Kent SBH (1986) Science 231:134-139
DeGrado WF, Kaiser ET (1980) J Org Chem 45:1295-1300
DeGrado WF, Kaiser ET (1982) J Org Chem 47:3258-3261
Fitch WM (1977) Genetics 86:623-644
Fukushima D, Kupferberg JP, Yokoyama S, Kroon DJ, Kaiser ET, Kezdy FJ (1979) J Am Chem Soc 101:3703-3704
Kaiser ET, Kezdy FJ (1984) Science 223:249-255
Mihara K, Kaiser ET (1988) Science, in press
Nakagawa SH, Kaiser ET (1983) J Org Chem 48:678-685
Nakagawa SH, Kaiser ET (1985) J Am Chem Soc 107:7087-7092
Sasaki T, Findeis M, personal communication

Computer Modeling with a Protein Data Base

T. A. Jones [1]

Introduction

The widespread use of computer graphics in the last 10 years has contributed to the spread and ease with which crystallographers can solve and refine macromolecular structures. In particular computer graphics modeling programs have been used to construct the initial model in electron density maps, and to remove model errors during crystallographic refinement (see Jones 1987, for a short review). In much of this interactive work, the experience of the modeler has played an important (sometimes critical) role. This manifests itself in an ability to recognize and use structural features. An intimate knowledge of protein stereochemistry can be of great use in constructing and modifying a known structure. The modeling program, Frodo (Jones 1978,1982) provided a set of tools to construct and manipulate fragments into maps in a straightforward general way. These fragments could be α-helices, β-strands or various types of turns, for example. In our recent work (Jones and Thirup 1986) we have generalized these concepts so that the fragments come directly from a data base of well refined protein structures.

Crystallographic Applications

The most frequently used representation of electron density has been "chicken wire" contouring where the 3-D map is contoured in three roughly perpendicular planes (usually the crystallographic planes). This representation has the advantage of producing a volume representation containing much detailed information. It has the major disadvantage of limiting the possible viewing volume simply because of the mass of information. Other representations have been suggested, in particular skeletonized density (Greer 1974). A computer graphics implementation of a similar representation has been made at Chapel Hill by Pique and co-workers (Pique 1984).

Our implementation (Jones and Thirup 1986) allows the display of both (or either) representations and offers the user the advantages of each sort. The reduced information in the skeleton allows one to easily view a much larger volume and offers a natural framework on which to test one's ideas of how a new structure is folded. By coloring the skeleton one can form a hypothesis of the structure that is easily saved (or restored). Since the normal MIR map contains many errors, it is necessary to modify the connectivity of the initial skeleton. This is usually done after inspecting the detailed contoured map.

[1]Department of Molecular Biology, University of Uppsala, BMC, Box 590, S-751 24 Uppsala, Sweden

39. Colloquium Mosbach 1988
Protein Structure and Protein Engineering

After editing the skeleton to represent the fold of the main chain together with branching side chains, it then becomes necessary to build a complete atom model with the desired sequence. This can be done by matching guide points on the skeleton to the protein data base. Normally one searches for a fragment of five to six amino acids. The side chains are added according to the desired sequence, in an extended conformation. These need to be fitted independently, either automatically (to the skeleton, or to the real density) or manually.

In Uppsala we have solved three structures in this way. Two rubisco structures have been solved by Brändéns group (Schneider et al. 1986) and P2 myelin protein by Jones et al. (1988). The rubisco structures contain ∿1000 amino acids and the P2 structure 400 amino acids in the asymmetric unit. It is of interest to describe the solution of P2 in more detail. The structure was solved with SIR/anomalous data from a single excellent derivative. The protein consists of 131 residues with three molecules in the asymmetric unit. Four different maps were calculated (SIR, two anomalous and a density modified SIR map) and the skeletons (at 1.5 the standard deviation of the map) of the complete unit cell displayed. One anomalous map clearly showed three protein like regions. One of these was interpreted in ∿3 h. This was carefully edited the following day, and the structure built and fitted the next day. This model was real space fitted (Jones and Liljas 1984) as a rigid body into the density of the other two molecules. The complete asymmetric unit was then real space fitted, splitting each residue into smaller fragments (Jones and Liljas 1984). In our future implementations we plan to completely automate this procedure from the stage when the user has decided on an hypothesis. The observation of Ponder and Richards (1987), that side chains can be constructed from a small number of rotamers, can easily be incorporated into this scheme. These were used in my P2 work so that whenever the side chain density was unclear, I used the most likely rotamer conformation.

The rapid developments in computer technology mean that searching for errors at the display is the time-limiting step in a crystallographic refinement study. To solve this problem we are concentrating our efforts on locating possible errors and giving a likely solution. It is then up to the user to accept this advice or not. Our initial efforts to locate peptide plane errors use the data base to find the best fitting set of fragments to each five-residue region and then to monitor the fit of the central carbonyl oxygen to those in the data base. It has been previously observed (Jones 1982) that reciprocal space refinement programs cannot flip incorrect peptides. They can, however, place the C_α and side chain atoms correctly even if the peptide plane is wrong. Our initial results suggest that these errors can be located and corrected. Similar programs to locate side chain errors can be developed. The crystallographic density (or difference density) is also available for inspection.

NMR Application

We have recently described a program to pick out fragments that satisfy NOE distance restraints (Kraulis and Jones 1987). We plan to use similar ideas to make local corrections to models produced by other methods.

Modeling Homologous Proteins

Jones and Thirup (1986) suggested that the data base could be useful in modeling a protein having sequence homology to a known structure. A blind test on α-lactalbumin starting from lysozyme coordinates showed that the correct fragments could be found to model deletions (r.m.s. fits to Cα's < 0.5 Å) but they moved as rigid bodies so that after comparison with the X-ray structure (kindly supplied by Dr. David Stuart and co-workers) the r.m.s. fit was high (2-3 Å). It remains to be seen if a dynamics/minimization system using main chain fragments and side chain rotamers could make the desired improvements.

Future Developments

Our laboratory has a computer network to which is attached compute and disk servers, and graphics workstations of different capabilities. The network also contains plotters and Postscript printers. In the immediate future we expect to dedicate a work station if not to every desk, at least to every project. These work stations are of different capabilities and will require resources from other units on the network. It will be important to make this connection as transparent as possible. Michael Levitt and I have recently made a "loosely coupled connection" between Frodo and external systems that makes it easy to acess the "world" from a work station. For a network solution, the adoption of a set of standards (X-windows, Postscript, PHIGS, TCP-IP) is critical.

References

Greer J (1974) Three-dimensional pattern recognition: An approach to automated interpretation of electron density maps of proteins. J Mol Biol 82:279-301

Jones TA (1978) A graphics model building and refinement system for macromolecules. J Appl Crystallogr 11:268-272

Jones TA (1982) FRODO: a graphics fitting program for macromolecules. In: Sayre D (ed) Computational crystallography, Oxford University Press, pp 303-317

Jones TA (1987) Electron density fitting with computer graphics: A review and a glimpse. In: Moras D, Drenth J, Strandberg B, Suck D, Wilson K (eds) Crystallography in molecular biology, Plenum Press, pp 125-130

Jones TA, Liljas L (1984) Crystallographic refinement of macromolecules having non-crystallographic symmetry. Acta Crystallogr A40:50-57

Jones TA, Thirup S (1986) Using known substructures in protein model building and crystallography. EMBO J 5:819-822

Jones TA, Bergfors T, Sedzik J, Unge T (1988) The three-dimensional structure of P2 myelin protein. EMBO J 7:1597-1604

Kraulis PJ, Jones TA (1987) Determination of three-dimensional protein structures from nuclear magnetic resonance data using fragments of known structures. Proteins: Structure, Function, and Genetics 2:188-201

Pique M (1984) Work in progress at UNC Chapel Hill NIH research resource. J Mol Graphics 2:59

Ponder JW, Richards FM (1987) Tertiary templates for proteins: use of packing criteria in the enumeration of allowed sequences for different structural classes. J Mol Biol 4:775-792

Schneider G, Lindqvist Y, Brändén C-I, Lorimer G (1986) Three-dimensional structure of ribulose-1,5-bisphosphate carboxylase/oxygenase from *rhodospirillum rubrum* at 2.9 Å resolution. EMBO J 5:3409-3415

Automated and High-Speed DNA Sequencing – Computer Technology Promotes Biological Advances

A. Wada [1]

Introduction

Forty years after the first electronic computer was developed, the use of this highly intelligent and versatile tool for studying molecular mechanisms of life has now become commonplace. Much of the impetus for the remarkable developments that have recently occurred, especially in the molecular aspect of biological science, is due to the vastly improved performance realized by the extensive application of computer and mechatronics to instrumentation science and technology. These developments are able to provide an extraordinary range of sensitivity, resolving power, dynamic range, and applicability, as well as to develop novel tools; only with the capability of present-day computers can complex structures and functions of living organisms be fully analyzed. In this chapter, I will show an aspect of the expanding applicability of modern technology including automation and robotics to the study of biological macromolecules with special reference to a large-scale DNA-base-sequencing system.

Information Flow in Biophysical Study

As everyone knows, a living organism contains, even in its single molecular units, i.e., DNA, RNA, or protein, a tremendous number of variables which represent both their structural and functional features. Thus molecular biophysical research ought to treat and process a large amount of complex signals to obtain refined final models.

In general, studying biological macromolecules is a series of operations of sample preparation, physical and chemical measurements, instrumentation technology, and information processing. Figure 1 shows the entire flow of materials, signals, and information in the course of molecular biophysical study from a starting material to the final model building and the input of data into a database.

In early days, the applications of the computer and related technologies were mostly restricted to the operation of and data processing in a physical instrument. During the past decade, however, the rapid development of inexpensive microcomputers and microprocessors, handy actuators, adaptable sensors, and a variety of display devices, as well as highly sophisticated algorithms for their operation and data analysis, have made possible large information-density interactive conversation between researchers and instruments. As a result, the computer technology application is now expanding in both up- and down-

[1]Department of Physics, Faculty of Science, The University of Tokyo, Bunkyo-ku, Tokyo, Japan

39. Colloquium Mosbach 1988
Protein Structure and Protein Engineering

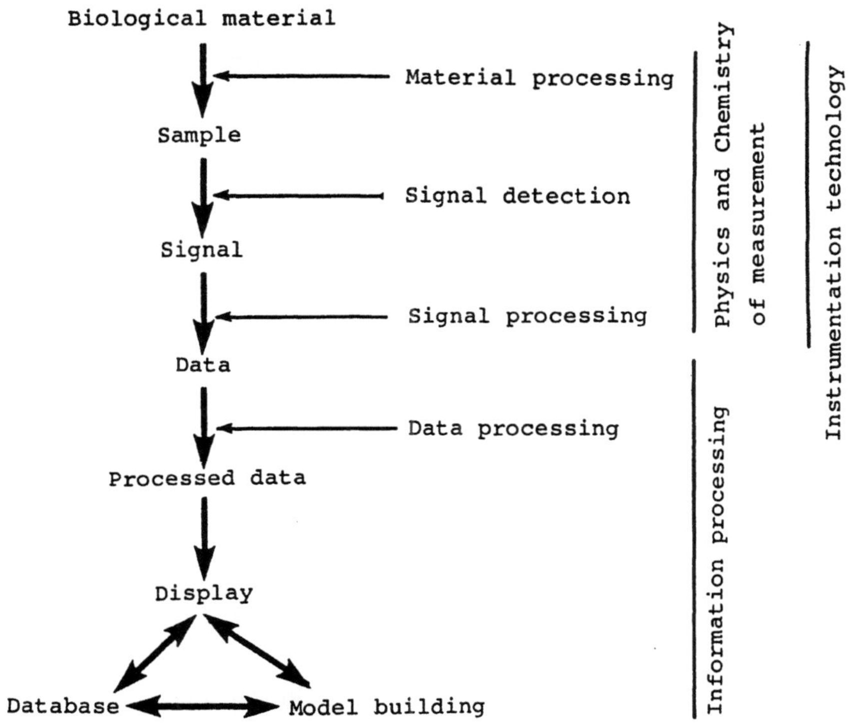

Fig. 1. Material, data, and information flow in biophysical measurement

stream directions to sample preparation and data processing in Fig.1, to meet the high information-extracting capability of measuring instruments.

The amount of information which can be extracted by a single experiment with a modern physical tool such as X-ray diffraction, electron microscopy, or NMR ranges from several kilos to megabytes, and rough estimations are given in Table 1. These figures seem to be the ultimate extent of the information accessible with today's single physical measurement.

By comparison, the amount of genetic information of a DNA is [2 bits × number of base pairs], which gives 6 Gbit in the case of the human genome DNA. This amount itself is already far beyond the capability of a single physical measurement, but it must be noted that for an accurate reading of the base sequence with one-base resolution, we must collect several hundred times more than this apparent amount of information simply estimated from the number of base pairs. Furthermore, the type of observables which could be accessed by a physical method alone to provide a large amount of information is highly limited; and DNA-base-sequence reading is just the sort of object which is hard to deal with, because it is hard to find effective differences among molecular physical properties of four bases in a DNA molecular strand. Thus a correlative effort should be made to reduce the signal to noise ratio to gain meaningful high grade data. In the DNA base sequence anal-

Table 1. Amount of information which is extracted with several high-resolution measurements

1. X-ray diffraction: No. of reflection points[a]

No. of atoms	1000	7000
Resolution		
6 Å	300	2000
3 Å	2500	18000
2 Å	7500	52000
1.5 Å	20000	140000

Information/point (8 bit) × no. of reflection

200 K ∿ 1 M bits

2. Three-dimensional image-reconstruction with electron microscopy: actin-tropomyosin-S1 at 15 Å resolution

Volume	Resolution	Information/data
9.6×10^7 Å^3	(1/7.5 Å^3)	10 bits = 2.3 M bits

3. NMR[c]

3.1. 1D NMR

No. of data points	Resolution
1.6×10^4	16 ∿ 32 bits = 200 ∿ 500 K bits

3.2 2D NMR

No. of data points	Resolution
4000 × 512	16 ∿ 32 bits = 33 ∿ 66 M bits

According to: [a]Dr. Yukio Mitsui (Univ. Tokyo), [b] Dr. Tateyuki Wakabayashi (Univ. Tokyo), and [c]Dr. Shigeru Endo (JEOL)

ysis, this noise reduction is performed by confining the measuring object to local sequences by fractionating them at the stage of the sample preparation, prior to the physical measurement.

DNA Base-Sequence Analysis

The essence of the base-sequence analysis is a series of specific or nonspecific fragmentations and precise fractionations of a sample DNA.

This biochemical cut-and-purify process is repeated until an optimum fragment length for the application of a physical measurement is obtained. The final measurement must be a physical one, because the sequence output should be electronic signals for the data processing and storage by a computer. The measurements are carried out on each of the fractionated fragments, and the whole sequence is eventually reconstructed from them by examining overlapping sequences in the fragments.

A typical example of the analysis of human genome, which has 3×10^9 base pairs (b) is as follows: A whole set of chromosomes is fractionated into 22 pairs plus X/Y components first by a cell sorter. The DNA strand in each chromosome thus separated is further subdivided spe-

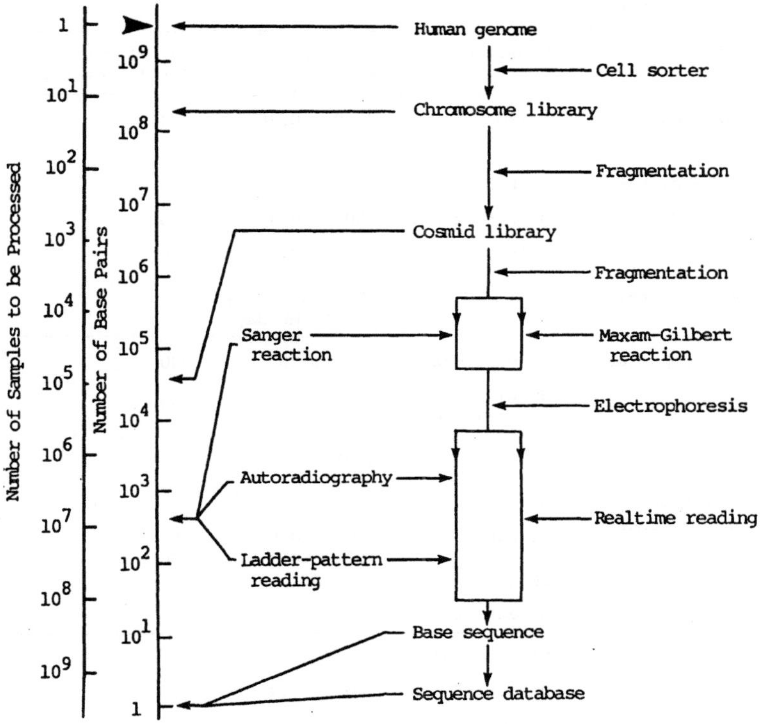

Fig. 2. Stages in the fragmentation and fractionation of a human genome: chromosome sorting, library construction, pre-electrophoresis, electrophoresis, and post-electrophoresis

cifically until an optimum size for the sequencing chemical reaction is reached (Fig.2).

In the course of this fragmentation, the number of bases in a fragment decreases from 3×10^9 to one, i.e., monomeric A, T, G, and C nucleotides, while the variety of fragments, i.e., the number of individual samples to be analyzed, increases from one at the beginning (one genome) to the order of 10^7 at electrophoresis, and every 3×10^9 letters should eventually be identified by the electrophoresis ladder pattern. This means that automated tools with high processing speed are especially required at the latter stages of analysis.

To be more quantitative, let us define the performance of a processing machine by the number of bases in samples which can be processed in one day – a unit of daily base-number output of a machine = b/D – or by the amount of time required for processing one base: s/b. 1 s/b corresponds to 86 Kb/D.

In designing a DNA sequencing system of Xb/D output consisting of a series of processing machines from the up- to down-stream of the analysis, the most appropriate arrangement of the units is given under the condition that the number of units i and j, N_i and N_j, are given by

$$Xbp/D = N_i\ (b/D)_i = N_j\ (b/D)_j = - - - ,$$

Table 2. Performance of sequencing machines

	Process	Developer	Performance Kb/D : s/b	No. of machines for 1 Mb/D output
Pre-electrophoresis	Purification of DNA	Seiko	300[a] : 0.28	4
	Maxam-Gilbert	Seiko	5 : 17	200
	M13/dideoxy	Martin et al.	220 : 0.4	5
		Seiko	300 : 0.28	4
	Gel preparation	Fuji	1000 films/day ∿1 Mb/day	1
Electrophoresis	Electrophoresis	Conventional	3 : 28	340
		Special performance	30[a] : 2.8	34
	Direct blotting	Pohl	(30)	
	Real time fluorescence detection	Smith et al.		
		Du Pont	10 : 8.6 ∿20 ∿4.3	100 ∿50
		EMBL		
		Hitachi-C		
Post-electrophoresis	Gel reader	Elder et al.		
		Hitachi-S	60 : 1.4	17
		Seiko		

[a]Expected speed.

Abbreviation: Seiko: Seiko Electronic and Instruments Co.; Martin: William J. Martin, The University of Manchester of Science and Technology; Fuji: Fuji Photo Film Co.; Pohl: F. Pohl, University of Konstanz; Smith: California Institute of Technology; Elder: J.K. Elder, University of Oxford; Hitachi-S.: Hitachi Software Co.; Hitachi-C: Hitachi Central Research Laboratory; EMBL: European Molecular Biology Laboratory; Du Pont: Du Pont de Nemoures & Co.

where $(b/D)_i$ and $(b/D)_j$ are the processing speeds of machines i and j, respectively. This equation indicates the output-input matching between machines, and sample flow conservation (i.e., no congestion of samples) in the process.

A continuous flow of a sample for the ceaseless operation of machines is essential for the efficiency and economy of the analysis: slow processes must be operated in parallel.

Table 2 shows a list of some machines which several groups in the world have developed or are developing. In the last column, the expected number of machines required for a 1 Megabase/D system.

A Large-Scale DNA Sequencing System: Its Practicability and Future Prospects

Table 2 indicates that a large-scale DNA-sequencing system able to achieve a daily output of one million bases (the 1 Mb/D super sequencer) can be realized by organizing current technology according to the principles of automation, robotics, and mass production line. The advantage of a single large sequencing center over small collecting stations for large-scale DNA sequencing is evident not only from the viewpoint of economy of labor and cost, but also of quality control to guarantee good reproducibility and high confidence in the results. All these advantages of a carefully organized mass processing system, including improved worker morale, have already been demonstrated in the production lines of modern industries from integrated circuits, TV's, radios, watches, to automobiles.

One might consider whether many "kilobase/day machines" or a few "megabase/day stations" should be developed. However, the problem will not be to make such an alternative judgement, but rather it is a matter of specialization. The situation is just like the sharing of computation by many desk-top calculators or a relatively few large or super computers. Thus, both Mb systems for large-scale research and Kb devices for laboratories should be developed to meet the researcher's demands.

Hopefully in the near future, extensive DNA-research centers equipped with a super DNA sequencer will be established at least in Europe, USA, and Japan as centers for international research cooperation open to researchers worldwide. In this regard, these centers should be connected to a worldwide sample-accepting and data-informing network, as well as to international DNA databases.

The significance of this undertaking should not be measured only by the current demand for such a system but should be evaluated in terms of the attraction of a proposed sequencing center to individuals and laboratories worldwide. A convenient and beneficial center would attract many samples, thus increasing the number of regular customers and in turn, increasing the efficiency of analysis and lowering of sequencing cost – the arrangement would be self-stimulating. Priority of the analysis may be discussed in an international organizing committee. Its focus should be placed first on health research to aid the finding of cures for such life-threatening human diseases as cancer, AIDS, and genetic diseases, as well as on the advancement of the basic biological sciences.

The need for a high performance sequencing system is now apparent and the time is ripe for its development due to the rapid pace of progress in related fields. There is increasingly widespread appreciation of the value of DNA databases and the enrichment of genetic libraries, especially of the human genome.

DNA sequence analysis, which now requires long, tedious, and repetitive manual work, is undeniably an appropriate subject for automation. The automation of this type of laboratory procedures could well turn out to be the equivalent of the Industrial Revolution in its effects on biological experimentation. The history of science and technology clearly indicates that the natural direction of progress is toward the extensive involvement of computer technology, including automation and robotics. It is hoped that the automated DNA sequencing system demonstrates a typical feature of the usefulness of modern tools in biology. In the 21st century we foresee Super DNA Sequencing Center taking its place beside high-energy particle accelerators, giant telescopes, and far-reaching space programs, as a symbol of human effort to broaden and build up the collective knowledge base of mankind.

Synthetic Antibodies with a Known Three-Dimensional Structure

A. Plückthun, A. Skerra, R. Glockshuber and J. Stadlmüller[1]

The problem of discriminating self from nonself with exquisite specificity has prompted nature to develop several classes of binding proteins, of which the antibodies are the best studied. Immunoglobulins are a family of rather stable and similar molecules that are able to bind to a large number of different antigens. It is therefore tempting to carry nature's system of "protein engineering" to the laboratory.

Antibodies constitute promising targets for investigating protein-ligand interactions, since the overall folding of the domains seems to be quite independent of the structure of the binding site. Several three-dimensional structures of antibody F_{ab} fragments have been determined and their common features have been compared (Kabat 1978; Amzel and Poljak 1979; Davies and Metzger 1983). More recently, structures of complexes consisting of F_{ab} fragment and protein antigen (Amit et al. 1986; Colman et al. 1987; Sheriff et al. 1987) were solved and have brought to light again the question of conformational changes in the antigen and antibody (Marquart and Deisenhofer 1982; Huber and Bennett 1987). The common structural feature of all antibody domains is a sandwich of antiparallel β-sheets. The architecture of the variable domains is characterized by a framework of fairly constant residues (arranged in the β-sheets) linked by three hypervariable loops (complementarity determining regions, CDR) per chain. These hypervariable loops contain the residues that make contact with the antigen.

Since the framework structure is fairly well conserved between antibodies, the loop conformations as the decisive factor in determining the specificity have been the subject of theoretical studies. Several approaches to the prediction of loop conformations have been taken, including conformational search procedures combined with energy minimization or molecular dynamics (Chothia et al. 1986; Fine et al. 1986; Shenkin et al. 1987; Bruccoleri and Karplus 1987), and structural homology searches (de la Paz et al. 1986; Chothia et al. 1986; T.A. Jones and Thirup 1986). The powers and limitations of these methods have not yet clearly emerged, particularly since few comparisons between their results and experimental data have been carried out. Clearly, there is an urgent need for a sufficiently large database of experimentally determined structures. The systematic variation of single parameters (such as the length or composition of single loops) and the modification of one single antibody may provide a deeper insight into the problem. Thus, an experimental system that permits the facile alteration of the antibody sequence would be of great value.

We have approached this challenge by using as a model system the exceptionally well-characterized phosphorylcholine binding myeloma protein McPC603, an immunoglobulin A of the mouse. Previous studies of this antibody include the determination of the sequence (Rudikoff and Potter 1974; Rudikoff et al. 1981), hapten binding constants (Metzger

[1]Genzentrum der Universität München, c/o Max-Planck-Institut für Biochemie, Am Klopferspitz, D-8033 Martinsried bei München, FRG

39. Colloquium Mosbach 1988
Protein Structure and Protein Engineering

Fig. 1. Ribbon diagram of the F_v fragment of McPC603. The coordinates of Padlan et al. (entry 2MCP in the Protein Data Bank) were used with the plot programm of Lesk and Hardman (1982). The α-carbon tracing of the light chain is colored *grey* and that of the heavy chain is shown in *white*. The hapten phosphorylcholine is colored according to atom types with O in *black*, P in *light grey*, C in *white* and N in *dark grey*

et al. 1971; Leon and Young 1971; Young and Leon 1977), binding kinetics (Goetze and Richards 1977) and, most importantly, the three-dimensional structure of the F_{ab} fragment (Segal et al. 1974; Satow et al. 1986). Furthermore, sequences of several related antibodies have been determined (Potter 1977; Perlmutter et al. 1984; Kabat et al. 1987) and the interactions between the variable domains have been analyzed (Chothia et al. 1985).

The bottleneck for any protein engineering studies on antibodies was a suitable expression system. Expression of functional whole antibodies has only been reported for cells of higher eukaryotes (Morrison and Oi 1984; Morrison 1985; Roberts and Rees 1986). Yet, genetic manipulations, efficient transformations, fast growth, and simple fermentation can all be carried out more easily in bacterial systems. Expression of a native antibody or binding fragment had, however, never been reported for any bacterial system. Previous attempts have been reviewed (Morrison 1985; Boss and Wood 1985). Antibody proteins were only obtained from inclusion bodies in *E. coli* (Cabilly et al. 1984; Boss et al. 1984), but the renaturation yielded only a small amount of functional protein. Similarly, only a fraction of the antibody protein produced by yeast was functional (Wood et al. 1985). Active antibody or an antigen binding fragment had not been purified from either yeast or *E. coli*.

We solved this problem by developing an expression system (Skerra and Plückthun 1988) that was designed to reproduce in *E. coli* the assembly process of antibodies in the eukaryotic cell. There, the immunoglobulin chains are secreted into the lumen of the endoplasmic reticulum (ER), where protein folding, assembly of the light and heavy chains and disulfide-bond formation occurs. In the lumen of the ER and subsequently in the Golgi apparatus the protein is also glycosylated. Our working

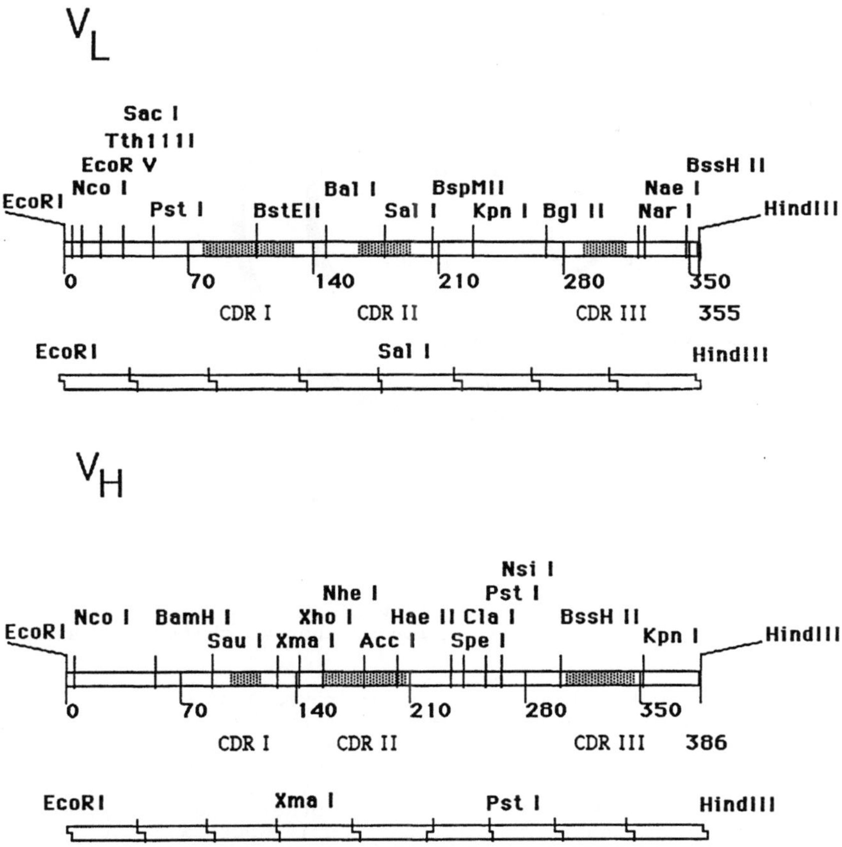

Fig. 2. Schematic representation of the synthetic genes. Unique restriction sites are indicated by a *vertical line with the label above it*. The complementarity determining regions *(CDR)* are *stippled*. The bottom representation of each gene schematically shows the oligonucleotides from which it was assembled, as well as the cloning sites used to assemble the gene from subclones (two subclones for V_L, 3 for V_H)

hypothesis was that the transport to the periplasm of *E. coli* is functionally equivalent to the transport to the ER in the eukaryotic cell. While the transport of heterologous proteins to the periplasm of *E. coli* with correct folding and correct S-S bond formation had been observed in several instances (for a review, see e.g., Briggs and Gierasch 1986) it was unknown whether two different proteins (light and heavy chain) would assemble to a heterodimer in *E. coli* in vivo. It is, for instance, not clear whether other proteins participate in the antibody assembly process in eukaryotes (Hendershot et al. 1987) and whether bacteria possess functionally equivalent proteins. A suitable bacterial expression system for functional antibody fragments must ensure (i) synthesis of approximately stoichiometric amounts of the two antibody chains, (ii) transport of both precursor proteins to the periplasmic space, (iii) correct processing of both signal sequences, resulting in the same N-termini as in the protein isolated from the mouse, (iv) folding to

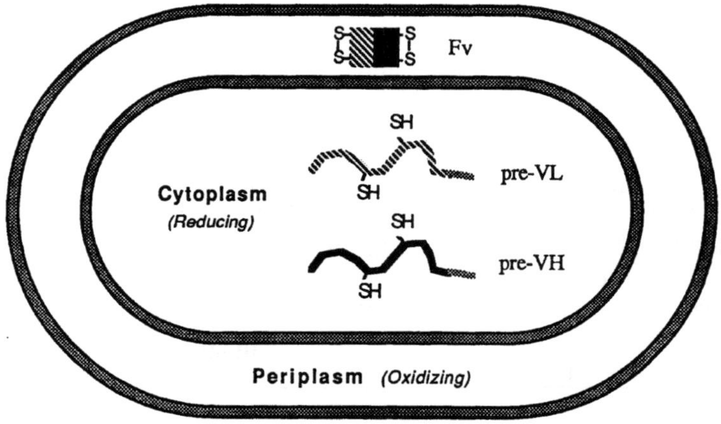

Fig. 3. Schematic diagram of the *E. coli* secretion system. In the cytoplasm, the precursor proteins for V_L and V_H, each fused to a bacterial signal sequence, are synthesized in reduced form. After translocation through the inner membrane into the periplasm, the signal sequences are cleaved off, the disulfide bonds form, and the domains fold and assemble (see text)

globular and soluble domains, (v) formation of the intramolecular disulfide bonds and finally (vi) association of the two chains to form a heterodimer.

The experiments were carried out with the F_V fragment of McPC603, which is the heterodimer of the variable domains of both the light and the heavy chain, V_L (115 amino acids) and V_H (122 amino acids) (Fig.1). Each domain contains one intramolecular disulfide bond (connecting cys 23 to cys 94 in V_L and connecting cys 22 to cys 98 in V_H). There is no disulfide bond between the chains and no other free sulfhydryl group. The genes for V_H and V_L were synthesized (Plückthun et al. 1987) (Fig.2) using the known protein sequence (Rudikoff and Potter 1974; Rudikoff et al. 1981). The gene sequences were designed to contain many unique restriction sites as well as not to contain codons very rare in *E. coli*. The two genes were fused to two different bacterial signal sequences and arranged in an artificial dicistronic operon, under the control of the inducible lac promoter. With this expression vector (Skerra and Plückthun 1988), both chains are secreted into the periplasmic space, where they assemble spontaneously to form a functional heterodimer (Fig.3). The recombinant F_V fragment so produced can be purified to homogeneity in a single step by affinity chromatography with the hapten phosphorylcholine as affinity ligand.

The functionality of the F_V fragment was tested by equilibrium dialysis with radioactive phosphorylcholine. These experiments showed that there is one binding site per F_V fragment, that there is no inactive protein in the preparation and that the affinity constant of the F_V fragment with phosphorylcholine ($1.21 \pm 0.06 \times 10^5$ M^{-1}) is, within experimental error, identical to that of intact McPC603 isolated from mouse ascites (Skerra and Plückthun 1988). It has been debated in the literature whether F_V fragments possess the same binding affinity as the native antibodies (Inbar et al. 1972; Sen and Beychok 1986), but our experiments showed normal affinity of the hapten to the F_V fragment of MCPC603. The main conclusion from these investigations is that the F_V fragment of McPC603 is a fully functional antibody model system for studying antigen-antibody interactions and that the very convenient *E. coli* expression system will facilitate antibody engineering.

We have also developed a second method for the expression of F_V fragments in *E. coli*. Fusion proteins between β-galactosidase and V_L or V_H were constructed in which the two proteins were linked by a recognition site for the highly specific blood clotting protease factor Xa. The denatured fusion proteins were obtained in large amounts from inclusion bodies (Plückthun et al. 1987). The F_V fragment was renatured in vitro from the combined cleavage reaction mixtures and purified by phosphorylcholine affinity chromatography (Glockshuber and Plückthun, in preparation). In the renatured F_V fragment both disulfide bonds were demonstrated to be present.

The use of bacteria as expression hosts for functional antibody fragments may open the door to new techniques that may be more difficult or impossible to achieve with other systems. The high transformation rate of *E. coli* should permit one to establish mutant libraries by using DNA that was subjected to random mutagenesis. Expression in the native state may allow the analysis of binding mutants directly on bacterial colonies. Even the use of positive metabolic selection for new functions may be envisaged, a potential very hard to realize in higher cells because of the paucity of selectable markers. In addition to the variable domains, the constant regions have been the focus of several recent investigations. For instance, marker enzymes (Neuberger et al. 1984), toxins (Möller 1982), constant regions from a different immunoglobulin class (Neuberger et al. 1985) or different species (P.T. Jones et al. 1986) are being used in experiments aimed at replacing the constant domains. The bacterial expression system may also facilitate the construction of such new hybrid antibodies.

To further understand the process of hetero-dimer assembly, we have investigated the properties of the separated domains V_L and V_H. For this separation, ion exchange chromatography in the presence of urea or reverse phase HPLC was used. The disulfide bonds remained intact during the separation procedure. It could be shown by FPLC size exclusion chromatography that V_L quantitatively dimerizes with itself (Glockshuber and Plückthun, in preparation). The sequence of the V_L domain is 63% identical to the sequence of the Bence-Jones protein REI (Palm and Hilschmann 1975), a V_L dimer from the urine of a multiple myeloma patient (Hobbs 1975). The three-dimensional structure of the REI dimer has been determined at high resolution (Epp et al. 1974,1975). The V_L domain interface is strongly conserved between a subunit of REI and V_L of McPC603. We propose, therefore, that the light chain of McPC603 dimerizes in a structure very similar to that found for REI. No indication was obtained for a dimerization of the V_H domain. A three-dimensional model for a putative V_H-dimer was generated from replacement of the V_L domain by V_H in the F_V-part of the structure of the F_{ab} fragment (Segal et al. 1974). The modeling suggested that such a structure is sterically not possible since the two CDR3 loops would collide. At high concentrations, V_H may form non-specific aggregates to aovid the exposure of the large hydrophobic inter-domain region to the solvent. It may be interesting in this context that it has been proposed that there is a specific protein in the eukaryotic cell that temporarily associates with the heavy chain (Hendershot et al. 1987).

The antibody McPC603 is also used by us as a model system for investigating enzymatic catalysis. Enzymes discriminate the transition state of the catalyzed reaction as the structure to be bound most tigthly.

By this strategy, part of the intrinsic binding energy of the ground state is converted into bringing the bound substrate closer to the transition state. The idea of transition state complementarity of enzymes was already considered by Haldane (1930) and Pauling (1946). Many transition state analogs have been synthesized and shown to be effective enzyme inhibitors (review: Wolfenden 1976). Recently, transition state analogs have also been used as haptens for immunization and were shown to indeed elicit catalytic antibodies (Raso and Stollar 1975a,b; Tramontano et al. 1986; Pollack et al. 1986; Jacobs et al. 1987).

A particularly attractive feature of McPC603 is that it binds a phosphate-ester and may therefore already be close in structure to a transition-state binding protein. It may thus stabilize the tetrahedral intermediate (and therefore, according to Hammonds principle, the transition states leading to and away from it) of, e.g., an ester hydrolysis of a properly designed substrate. The antibody may therefore act like an enzyme. Phosphates and phosphonates have been demonstrated to act as transition state analogs for enzymatic carboxyl-ester or amide hydrolysis reactions (Weaver et al. 1977; Kam et al. 1979; Hofmann and Rottenberg 1980; Jacobsen and Bartlett 1981; Thorsett et al. 1982; Bartlett and Marlowe 1983; Galardy et al. 1983; Tronrud et al. 1986; Bartlett and Marlowe 1987; Tronrud et al. 1987). Esterases and proteases have been found to be complementary in structure to the tetrahedral intermediate of the hydrolysis reaction (Kraut 1977). Recently, a serine protease that was stripped of its catalytic triad has been shown to display residual activity, presumably by transition state stabilization (Carter and Wells 1988). The antibody McPC603 is a very suitable system with which to investigate catalytic rate enhancements, since the structure of the complex with the hapten is known and since we have devised a convenient expression system that permits us to modify the protein (see above). With this system, theories on rate acceleration may therefore be directly tested.

While previous experiments with monoclonal catalytic antibodies have demonstrated the validity of Haldane's and Pauling's concepts, they have as yet provided little information on the nature of this catalysis, since structural information is available for none of these antibodies. In the case of McPC603, however, the three-dimensional structure with the hapten bound is known. Protein-engineering of this molecule should deepen our understanding of catalytic requirements through experiments aimed at modifying the putative active site.

References

Amit AG, Mariuzza RA, Phillips SEV, Poljak RJ (1986) Three-dimensional structure of an antigen-antibody complex at 2.8 Å resolution. Science 233:747-753

Amzel LM, Poljak RJ (1979) Three-dimensional structure of immunoglobulins. Annu Rev Biochem 48:961-997

Bartlett PA, Marlowe CK (1983) Phosphonamidates as transition state analogue inhibitors of thermolysin. Biochemistry 22:4618-4624

Bartlett PA, Marlowe CK (1987) Evaluation of intrinsic binding energy from a hydrogen bonding group in an enzyme inhibitor. Science 235:569-571

Boss MA, Wood CR (1985) Genetically engineered antibodies. Immunol Today 6:12-13

Boss MA, Kenten JH, Wood CR, Emtage JS (1984) Assembly of functional antibodies from immunoglobulin heavy and light chains synthesized in E.coli. Nucleic Acids Res 12:3791-3806

Briggs MS, Gierasch LM (1986) Molecular mechanisms of protein secretion: The role of the signal sequence. Adv Protein Chem 38:109-180

Bruccoleri RE, Karplus M (1987) Prediction of the folding of short polypeptide segments by uniform conformational sampling. Biopolymers 26:137-168

Cabilly S, Riggs AD, Pande H, Shively JE, Holmes WE, Rey M, Perry LJ, Wetzel R, Heyneker HL (1984) Generation of antibody activity from immunoglobulin polypeptide chains produced in *Escherichia coli*. Proc Natl Acad Sci USA 81:3273-3277

Carter P, Wells JA (1988) Dissecting the catalytic triad of a serine protease. Nature 332:564-568

Chothia C, Novotny J, Bruccoleri R, Karplus M (1985) Domain association in immunoglobulin molecules. J Mol Biol 186:651-663

Chothia C, Lesk AM, Levitt M, Amit AG, Mariuzza RA, Phillips SEV, Poljak RJ (1986) The predicted structure of immunoglobulin D1.3 and its comparison with the crystal structure. Science 233:755-758

Colman PM, Laver WG, Varghese JN, Baker AT, Tulloch PA, Air GM, Webster RG (1987) Three-dimensional structure of a complex of antibody with influenza virus neuraminidase. Nature 326:358-363

Davies DR, Metzger H (1983) Structural basis of antibody function. Annu Rev Immunol 1:87-117

de la Paz P, Sutton BJ, Darsley MJ, Rees AR (1986) Modelling of the combining sites of three anti-lysozyme monoclonal antibodies and of the complex between one of the antibodies and its epitope. EMBO J 5:415-425

Epp O, Colman P, Fehlhammer H, Bode W, Schiffer M, Huber R, Palm W (1974) Crystal and molecular structure of a dimer composed of the variable portions of the Bence-Jones protein REI. Eur J Biochem 45:513-524

Epp O, Lattmann EE, Schiffer M, Huber R, Palm W (1975) The molecular structure of a dimer composed of the variable portions of the Bence-Jones protein REI refined at 2.0-Å resolution. Biochemistry 14:4943-4952

Fine RM, Wang H, Shenkin PS, Yarmush DL, Levinthal C (1986) Predicting antibody hypervariable loop conformations II: Minimization and molecular dynamic studies of McPC603 from many randomly generated loop conformations. Proteins 1:342-362

Galardy RE, Kontoyiannidou-Ostrem V, Kortylewicz ZP (1983) Inhibition of angiotensin converting enzyme by phosphonic amides and phosphonic acids. Biochemistry 22: 1990-1995

Goetze AM, Richards JH (1977) Magnetic resonance studies of the binding site interactions between phosphorylcholine and specific mouse myeloma immunoglobulin. Biochemistry 16:228-232

Haldane JBS (1930) Enyzmes. Longmans, Green and Co., London; (1965) MIT Press, Cambridge, MA

Hendershot L, Bole D, Kearney JF (1987) The role of immunoglobulin heavy chain binding protein in immunoglobulin transport. Immunol Today 8:111-114

Hobbs JR (1975) Bence-Jones proteins. Essays Med Biochem 1:105-131

Hofmann W, Rottenberg M (1980) A transition state analogous organophosphate inhibitor of carboxypeptidase A. In: Brodbeck U (ed) Enzyme inhibitors. Verlag Chemie, Weinheim, pp 19-26

Huber R, Bennett WS (1987) Antibody-antigen flexibility. Nature (Lond) 326:334-335

Inbar D, Hochman J, Givol D (1972) Localization of antibody-combining sites within the variable portions of heavy and light chains. Proc Natl Acad Sci USA 69:2659-2662

Jacobs J, Schultz PG, Sugasawara R, Powell M (1987) Catalytic antibodies. J Am Chem Soc 109:2174-2176

Jacobsen NE, Bartlett PA (1981) A phosphonamidate dipeptide analogue as an inhibitor of carboxypeptidase A. J Am Chem Soc 103:654-657

Jones PT, Dear PH, Foote J, Neuberger MS, Winter G (1986) Replacing the complementarity-determining regions in a human antibody with those from a mouse. Nature 321:522-525

Jones TA, Thirup S (1986) Using known substructures in protein model building and crystallography. EMBO J 5:819-822

Kabat EA (1978) The structural basis of antibody complementarity. Adv Protein Chem 32:1-75

Kabat EA, Wu TT, Reid-Miller M, Perry HM, Gottesman K (1987) Sequences of proteins of immunological interest. 4th edn. US Dept of Health and Human Services. Public Health Service. Natl Inst of Health, Bethesda, Maryland
Kam CM, Nishino N, Powers JC (1979) Inhibition of thermolysin and carboxypeptidase A by phosphoramidates. Biochemistry 18:3032-3038
Kraut J (1977) Serine proteases: structure and mechanism of catalysis. Annu Rev Biochem 46:331-358
Leon MA, Young NM (1971) Specificity for phosphorylcholine of six murine myeloma proteins reactive with pneumococcus C polysaccharide and β-lipoprotein. Biochemistry 10:1424-1429
Lesk AM, Hardman KD (1982) Computer-generated schematic diagrams of protein structures. Science 216:539-540
Marquart M, Deisenhofer J (1982) The three-dimensional structure of antibodies. Immunol Today 3:160-166
Metzger H, Chesebro B, Hadler NM, Lee J, Otchin N (1971) Modification of immunoglobulin combining sites. In: Amos B (ed) Progress in immunology. Proceedings of the 1st Congress of Immunology. Academic Press, New York, pp 253-267
Möller G (ed) (1982) Antibody carriers of drugs and toxins in tumor therapy. Immunol Rev 62
Morrison SL (1985) Transfectomas provide novel chimeric antibodies. Science 229: 1202-1207
Morrison SL, Oi VT (1984) Transfer and expression of immunoglobulin genes. Annu Rev Immunol 2:239-256
Neuberger MS, Williams GT, Fox RO (1984) Recombinant antibodies possessing novel effector functions. Nature (Lond) 312:604-608
Neuberger MS, Williams GT, Mitchell EB, Jouhal SS, Flanagan JG, Rabbitts TH (1985) A hapten-specific chimaeric IgE antibody with human physiological effector function. Nature (Lond) 314:268-270
Palm W, Hilschmann N (1975) Die Primärstruktur einer kristallinen monoklonalen Immunglobulin-L-Kette vom κ-Typ, Subgruppe I (Bence-Jones-Protein Rei.) Hoppe Seyler's Z Physiol Chem 356:167-191
Pauling L (1946) Molecular architecture and biological reactions. Chem Eng News 24:1375-1377
Perlmutter RM, Crews ST, Douglas R, Sorensen G, Johnson N, Nivera N, Gearhart PJ, Hood L (1984) The generation of diversity in phosphorylcholine-binding antibodies. Adv Immunol 35:1-37
Plückthun A, Glockshuber R, Pfitzinger I, Skerra A, Stadlmüller J (1987) Engineering of antibodies with a known three-dimensional structure. Cold Spring Harbor Symp Quant Biol 52:105-112
Pollack SJ, Jacobs JW, Schultz PG (1986) Selective chemical catalysis by an antibody. Science 234:1570-1573
Potter M (1977) Antigen binding myeloma proteins of mice. Adv Immunol 25:141-211
Raso V, Stollar BD (1975a) The antibody-enzyme analogy. Characterization of antibodies to phosphopyridoxyltyrosine derivatives. Biochemistry 14:584-591
Raso V, Stollar BD (1975b) The antibody-enzyme analogy. Comparison of enzymes and antibodies specific for phosphopyridoxyltyrosine. Biochemistry 14:591-599
Roberts S, Rees AR (1986) The cloning and expression of an anti-peptide antibody: a system for rapid analysis of the binding properties of engineered antibodies. Protein Eng 1:59-65
Rudikoff S, Potter M (1974) Variable region sequence of the heavy chain from a phosphorylcholine binding myeloma protein. Biochemistry 13:4033-4038
Rudikoff S, Satow Y, Padlan E, Davies D, Potter M (1981) Kappa chain structure from a crystallized murine F_{ab}: role of joining segment in hapten binding. Mol Immunol 18:705-711
Satow Y, Cohen GH, Padlan A, Davies DR (1986) Phosphocholine binding immunoglobulin F_{ab} McPC603: An X-ray diffraction study at 2.7 Å. J Mol Biol 190:593-604
Segal DM, Padlan EA, Cohen GH, Rudikoff S, Potter M, Davies DR (1974) The three-dimensional structure of a phosphorylcholine-binding mouse immunoglobulin F_{ab} and the nature of the antigen binding site. Proc Natl Acad Sci USA 71:4298-4302
Sen J, Beychok S (1986) Proteolytic dissection of a hapten binding site. Proteins 1:256-262

Shenkin PS, Yarmush DL, Fine RM, Wang HJ, Levinthal C (1987) Predicting antibody hypervariable loop conformation I: Ensembles of random conformations for ring-like structures. Biopolymers 26:2053-2085
Sheriff S, Silverton EW, Padlan EA, Cohen GH, Smith-Gill SJ, Finzel BC, Davies DR (1987) Three-dimensional structure of an antibody-antigen complex. Proc Natl Acad Sci USA 84:8075-8079
Skerra A, Plückthun A (1988) Assembly of a functional immunoglobulin F_V fragment in Escherichia coli. Science 240:1038-1041
Snow ME, Amzel LM (1986) Calculating three-dimensional changes in protein structure due to amino acid substitutions: The variable region of immunoglobulins. Proteins 1:267-279
Thorsett ED, Harris EE, Peterson ER, Greenlee WJ, Patchett AA, Ulm EH, Vassil TC (1982) Phosphorus-containing inhibitors of angiotensin-converting enzyme. Proc Natl Acad Sci USA 79:2176-2180
Tramontano A, Janda KD, Lerner RA (1986) Catalytic antibodies. Science 234:1566-1570
Tronrud DE, Monzingo AF, Matthews BW (1986) Crystallographic structural analysis of phosphoramidates as inhibitors and transition-state analogs of thermolysin. Eur J Biochem 157:261-268
Tronrud DE, Holden HM, Matthews BW (1987) Structures of two thermolysin-inhibitor complexes that differ by a single hydrogen bond. Science 235:571-574
Weaver LH, Kester WR, Matthews BW (1977) A crystallographic study of the complex of phosphoramidon with thermolysin. A model for the presumed catalytic transition state and for the binding of extended substrates. J Mol Biol 114:119-132
Wolfenden R (1976) Transition state analog inhibitors and enzyme catalysis. Annu Rev Biophys Bioeng 5:271-306
Wood CR, Boss MA, Kenten JH, Calvert JE, Roberts NA, Emtage JS (1985) The synthesis and in vivo assembly of functional antibodies in yeast. Nature (Lond) 314:446-449
Young NM, Leon MA (1977) The binding of analogs of phosphorylcholine by the murine myeloma proteins McPC603, McPC167 and S107. Immunochemistry 14:757-761

MIX
Papier aus verantwortungsvollen Quellen
Paper from responsible sources
FSC® C105338

If you have any concerns about our products,
you can contact us on
ProductSafety@springernature.com

In case Publisher is established outside the EU,
the EU authorized representative is:
Springer Nature Customer Service Center GmbH
Europaplatz 3, 69115 Heidelberg, Germany

Printed by Libri Plureos GmbH
in Hamburg, Germany